NativeScript for Angular Mobile Development

Creating dynamic mobile apps for iOS and Android

Nathan Walker
Nathanael J. Anderson

BIRMINGHAM - MUMBAI

NativeScript for Angular Mobile Development

First published: August 2017

Production reference: 1280817

Published by Packt Publishing Ltd.
Livery Place
35 Livery Street
Birmingham
B3 2PB, UK.

ISBN 978-1-78712-576-6

www.packtpub.com

Credits

Authors
Nathan Walker
Nathanael J. Anderson

Reviewer
Alexander Vakrilov

Commissioning Editor
Amarabha Banerjee

Acquisition Editor
Prachi Bisht

Content Development Editor
Roshan Kumar

Technical Editor
Akhil Nair

Copy Editors
Akshata Lobo
Dhanya Baburaj

Project Coordinator
Devanshi Doshi

Proofreader
Safis Editing

Indexer
Mariammal Chettiyar

Graphics
Jason Monteiro

Production Coordinator
Shraddha Falebhai

Foreword

JavaScript developers faced a steep uphill battle if they wanted to write a mobile app. Not only did they have to learn completely new programming languages, they also had to deal with radically different development environments, as well as a dizzying set of testing and deployment procedures.

Today, NativeScript lets JavaScript developers write mobile apps from a single codebase using the language they already know and love--namely, JavaScript! By allowing you to use familiar frontend technologies, NativeScript drastically reduces the amount of time it takes you to write powerful and compelling iOS and Android applications.

However, it doesn't stop with the JavaScript language. NativeScript also lets you write apps with the Angular framework, allowing you to code with a frontend framework on platforms that used to make you learn Swift, Java, and Objective-C.

In this book, you'll learn how it all works from two of the NativeScript community's most talented developers.

Nathan Walker's list of contributions to the NativeScript world is too long for this foreword. Just to give you a taste though, Nathan Walker led the effort to create the NativeScript framework's first theme. Nathan also maintains the most popular project to share code between Angular web and native apps. However, in my personal opinion, Nathan's biggest contribution to NativeScript has been the tireless help he's offered to countless developers throughout the NativeScript community.

Nathan might only be outdone by Nathanael, who was one of NativeScript's first users and has been a regular in NativeScript's community chat and forums since day one. Nathanael Anderson understands NativeScript's inner workings better than most (all?) of the NativeScript team. He regularly contributes to all facets of the NativeScript framework, and he routinely shares his knowledge with the greater NativeScript world through his blog.

The two authors have the sort of real-world experience that you will need to cover these sort of topics in detail. Nathan can speak knowledgeably about the NativeScript core theme because he basically wrote the thing. Nathanael can talk about NativeScript unit testing approaches because he literally wrote the book on the topic.

I can't think of two better people to learn NativeScript and Angular from--you're in for a treat.

TJ VanToll
Principal Developer Advocate, Progress

About the Authors

Nathan Walker has enjoyed the opportunity to work in the web/mobile app development arena for more than 15 years. He cofounded nStudio LLC, a professional software development services and consulting company specializing in Angular and NativeScript integrations, while also working with Tryon Creek Software based out of Portland, OR. His varied background rooted in the world of design and arts provides him with a unique approach to problem solving. Spending several years working across multiple industries, including entertainment, audio/video production, manufacturing, b2b marketing, communications, and technology, helped establish an attitude that is focused on client needs.

I would like to thank the incredibly inspiring NativeScript and Angular communities for constantly striving to attain the best quality and most effective solutions for web and mobile development. In addition to the tremendous work put forth by both the core NativeScript and Angular teams, nothing would have been possible without their technical expertise and perseverance to produce quality software frameworks. I would also like to thank my coauthor, Nathanael Anderson, and Progress, along with their incredible Developer Relations group, including Jen Looper, TJ VanToll, Dan Wilson, Todd Anglin, Brian Rinaldi, Rob Lauer, and many more. An extra special thanks to Alex Vakrilov and Stanimira Vlaeva along with all the core team members for putting their blood, sweat, and tears into the amazing feat that is NativeScript for Angular. Lastly but most importantly, words alone cannot express my appreciation and thanks for the enduring love and support that my lovely wife, Kylie, provides me day in and day out, alongside the energy and motivation my wonderfully talented son Cole and extended family surround me with all throughout my life.

Nathanael J. Anderson has been developing software for more than 20 years in a wide range of industries, including games, time management, imaging, service, printing, accounting, land management, security, web and even, believe it or not, some actually successful government projects. He is currently the owner of Master Technology and cofounder of nStudio, LLC, and can create a solution for any type of application (native, web, mobile, and hybrid) running on any operating system. As a senior development engineer, he can work, tune, and secure everything from your backend servers to the final destination of the data on your desktop or mobile devices. By having a great understanding of the entire infrastructure, including the real or virtualized hardware, he can totally eliminate many different types of issue between all parts of the framework. He currently runs the entire NativeScript.rocks family of sites, has multiple highly rated cross-platform plugins for NativeScript, and works heavily for the NativeScript community.

Words fail to express how much I would like to thank my awesome wife Camarell for all her support, and my wonderful kids for the many times I was unavailable because I was coding or writing. In addition, I would like to thank the NativeScript developer and developer relations teams for the really cool framework that I have had the pleasure of using for several years since its release.

About the Reviewer

Alexander Vakrilov is one of the core contributors to the NativeScript project. He's got a love for hacking and optimizing things. In his spare time, he likes to jam on his guitar with friends and is always looking for opportunities for a pleasant conversation.

Thanks to the awesome team that made this book possible most notably Nathan and Nathanael for writing such great content. A special thanks goes to the whole NativeScript team for putting so much passion in the work we do. Most importantly, thanks to my lovely wife Irena and to my 2 year old son Vladi, who never understood why daddy is so sleepy after those nights of debugging.

www.PacktPub.com

For support files and downloads related to your book, please visit www.PacktPub.com. Did you know that Packt offers eBook versions of every book published, with PDF and ePub files available? You can upgrade to the eBook version at www.PacktPub.com and as a print book customer, you are entitled to a discount on the eBook copy. Get in touch with us at service@packtpub.com for more details.

At www.PacktPub.com, you can also read a collection of free technical articles, sign up for a range of free newsletters and receive exclusive discounts and offers on Packt books and eBooks.

https://www.packtpub.com/mapt

Get the most in-demand software skills with Mapt. Mapt gives you full access to all Packt books and video courses, as well as industry-leading tools to help you plan your personal development and advance your career.

Why subscribe?

- Fully searchable across every book published by Packt
- Copy and paste, print, and bookmark content
- On demand and accessible via a web browser

Customer Feedback

Thanks for purchasing this Packt book. At Packt, quality is at the heart of our editorial process. To help us improve, please leave us an honest review on this book's Amazon page at `https://www.amazon.com/dp/1787125769`.

If you'd like to join our team of regular reviewers, you can e-mail us at `customerreviews@packtpub.com`. We award our regular reviewers with free eBooks and videos in exchange for their valuable feedback. Help us be relentless in improving our products!

Table of Contents

Preface

NativeScript is an open source framework built by Progress to build truly native mobile apps with Angular, TypeScript, or even good old plain JavaScript. Angular is also an open source framework built by Google that offers declarative templates, dependency injection, and rich modules to build applications. Angular's versatile view handling architecture allows your views to be rendered as real native UI components--native to iOS or Android--that offer superior performance with fluid usability. This decoupling of the view rendering layer in Angular, combined with the power of native APIs in NativeScript, has come together to create the powerful combination that is the exciting world of NativeScript for Angular.

This book focuses on the key concepts you need to know to build NativeScript for your Angular mobile app on iOS and Android. We'll build a fun multitrack recording studio app, touching on the powerful native key concepts you need to know when you start building an app of your own. Having the right structure is critical to developing an app that can scale while also being highly maintainable and portable, so we'll start with project organization using Angular's @NgModule. We'll use Angular Components to build our first view and then create services that we can use via Angular's dependency injection.

You'll understand NativeScript's tns command-line utility to run the app on iOS and Android. We'll integrate third-party plugins to construct some of the core features. Next, we'll integrate the @ngrx store plus effects to establish some solid practices (Redux inspired) to deal with state management. Having a great data flow and solid architecture is meaningless if the app doesn't look good or offer a great user experience, so we'll use SASS to polish a style for our app. After that, we'll deal with debugging problems and invest some time into writing tests to prevent bugs in the future. Lastly, we'll bundle our app with webpack and deploy it to the Apple App Store and Google Play.

By the end of the book, you'll know the majority of the key concepts needed to build a NativeScript for Angular app.

What this book covers

`Chapter 1`, *Get into Shape with @NgModule*, discusses the @NgModule decorator, which clearly defines a segment of functionality in your app . This will be the organizational unit of your project. Before you begin building your app, you will gain many benefits by taking a moment and thinking about the various units/sections/modules that you may need/want for your app.

`Chapter 2`, *Feature Modules*, teaches you that structuring your app with feature modules offers many advantages for maintainability in the future and reduces duplication of code throughout your app.

`Chapter 3`, *Our First View via Component Building*, actually lets us see our app for the first time, where we need to build a Component for our first view.

`Chapter 4`, *A prettier view with CSS*, looks at how to turn our first view into something pretty amazing with a few CSS classes. We will also focus on how to utilize NativeScript's core theme to provide a consistent styling framework to build on.

`Chapter 5`, *Routing and Lazy Loading*, allows users to navigate around the various views in our app that will need to set up routing. Angular provides a powerful router that, when combined with NativeScript, works hand in hand with the native mobile page navigation system on iOS and Android. Additionally, we will set up the lazy loading of various routes to ensure that our app's launch time remains as speedy as possible.

`Chapter 6`, *Running the App on iOS and Android*, focuses on how to run our app on iOS and Android via NativeScript's tns command-line utility.

`Chapter 7`, *Building the Multitrack Player*, covers plugin integration and provides a direct access to Objective C/Swift APIs on iOS and Java APIs on Android via NativeScript.

`Chapter 8`, *Building an Audio Recorder*, works with native APIs to build an audio recorder for iOS and Android.

`Chapter 9`, *Empowering your views*, takes advantage of Angular's flexibility and NativeScript's power to get the most out of your app's user interface.

`Chapter 10`, *@ngrx/store + @ngrx/effects for state management*, manages app state via a single store with ngrx.

`Chapter 11`, *Polish with SASS*, integrates the nativescript-dev-sass plugin to polish our app's style with SASS.

`Chapter 12`, *Unit testing*, set up the Karma unit testing framework to future proof our app.

`Chapter 13`, *Integration Testing with Appium*, sets up Appium for integration testing.

`Chapter 14`, *Deployment Preparation with webpack Bundling*, works with webpack to optimize the bundle for release.

`Chapter 15`, *Deploying to the Apple App Store*, lets us distribute our app via the Apple App Store.

`Chapter 16`, *Deploying to Google Play*, lets us distribute our app via Google Play.

What you need for this book

This book assumes that you are using NativeScript 3 or higher and Angular 4.1 or higher. If you plan to follow along for iOS development, you will need a Mac with XCode installed to run the accompanying app. You should also have the Android SDK tools installed with at least one emulator, preferably running 7.0.0 with API 24 or higher.

Who this book is for

This book is for all types of software developer who are interested in mobile app development for iOS and Android. It's specifically tailored to benefit those who already have a general understanding of TypeScript and some basic-level Angular features. Web developers who are just getting into mobile app development for iOS and Android may also gain a lot from the content in this book.

Conventions

In this book, you will find a number of text styles that distinguish between different kinds of information. Here are some examples of these styles and an explanation of their meaning.

Code words in text, database table names, folder names, filenames, file extensions, pathnames, dummy URLs, user input, and Twitter handles are shown as follows: "Various common properties (`padding`, `font size`, `font weight`, `color`, `background color`, and more) are supported. Also, shorthand margin/padding works as well, that is, padding: 15 5."

A block of code is set as follows:

```
[default]
export class AppComponent {}
```

When we wish to draw your attention to a particular part of a code block, the relevant lines or items are set in bold:

```
[default]
public init() {
 const item = {};
 item.volume = 1;
 }
```

Any command-line input or output is written as follows:

```
# tns run ios --emulator
```

New terms and **important words** are shown in bold. Words that you see on the screen, for example, in menus or dialog boxes, appear in the text like this: "Running our app again, we now see the login prompt when we tap the **Record** button".

Warnings or important notes appear like this.

Tips and tricks appear like this.

Reader feedback

Feedback from our readers is always welcome. Let us know what you think about this book-what you liked or disliked. Reader feedback is important for us as it helps us develop titles that you will really get the most out of. To send us general feedback, simply email `feedback@packtpub.com`, and mention the book's title in the subject of your message. If there is a topic that you have expertise in and you are interested in either writing or contributing to a book, see our author guide at `www.packtpub.com/authors`.

Customer support

Now that you are the proud owner of a Packt book, we have a number of things to help you to get the most from your purchase.

Downloading the example code

You can download the example code files for this book from your account at `http://www.packtpub.com`. If you purchased this book elsewhere, you can visit `http://www.packtpub.com/support` and register to have the files emailed directly to you. You can download the code files by following these steps:

1. Log in or register to our website using your email address and password.
2. Hover the mouse pointer on the **SUPPORT** tab at the top.

3. Click on **Code Downloads & Errata**.
4. Enter the name of the book in the **Search** box.
5. Select the book for which you're looking to download the code files.
6. Choose from the drop-down menu where you purchased this book from.
7. Click on **Code Download**.

Once the file is downloaded, please make sure that you unzip or extract the folder using the latest version of:

- WinRAR / 7-Zip for Windows
- Zipeg / iZip / UnRarX for Mac
- 7-Zip / PeaZip for Linux

The code bundle for the book is also hosted on GitHub at `https://github.com/PacktPublishing/NativeScript-for-Angular-Mobile-Development`. We also have other code bundles from our rich catalog of books and videos available at `https://github.com/PacktPublishing/`. Check them out!

Downloading the color images of this book

We also provide you with a PDF file that has color images of the screenshots/diagrams used in this book. The color images will help you better understand the changes in the output. You can download this file from `https://www.packtpub.com/sites/default/files/downloads/NativeScriptforAngularMobileDevelopment_ColorImages.pdf`.

Errata

Although we have taken every care to ensure the accuracy of our content, mistakes do happen. If you find a mistake in one of our books-maybe a mistake in the text or the code-we would be grateful if you could report this to us. By doing so, you can save other readers from frustration and help us improve subsequent versions of this book. If you find any errata, please report them by visiting `http://www.packtpub.com/submit-errata`, selecting your book, clicking on the **Errata Submission Form** link, and entering the details of your errata. Once your errata are verified, your submission will be accepted and the errata will be uploaded to our website or added to any list of existing errata under the Errata section of that title. To view the previously submitted errata, go to `https://www.packtpub.com/books/content/support` and enter the name of the book in the search field. The required information will appear under the **Errata** section.

Piracy

Piracy of copyrighted material on the internet is an ongoing problem across all media. At Packt, we take the protection of our copyright and licenses very seriously. If you come across any illegal copies of our works in any form on the internet, please provide us with the location address or website name immediately so that we can pursue a remedy. Please contact us at `copyright@packtpub.com` with a link to the suspected pirated material. We appreciate your help in protecting our authors and our ability to bring you valuable content.

Questions

If you have a problem with any aspect of this book, you can contact us at `questions@packtpub.com`, and we will do our best to address the problem.

1
Get Into Shape with @NgModule

In this chapter, we are going to kick things off with some solid project organization exercises to prepare us for building an amazing app with NativeScript for Angular. We want to give you some insights into a few important and powerful concepts to think about as you plan your architecture, to pave your way to a smooth development experience with scalability in mind.

Combining Angular with NativeScript provides a wealth of useful paradigms and tools to construct and plan your app. As often said, with great power comes great responsibility, and as awesome as this tech combination is to create amazing apps, they can also be used for creating an overengineered and difficult to debug app. Let's take a few chapters to walk through some exercises you can use to help avoid common pitfalls and truly unlock the full potential of this stack.

We will introduce you to Angular's `@NgModule` decorator, which we will use exclusively to help organize our app's code into logical units with explicit purpose and portability in mind. We will also introduce a few Angular concepts we will use in our architecture, such as dependency injectable services. After doing our diligence in building a solid foundation to work with, we will rapidly approach running our app for the first time towards the end of third chapter.

In this chapter, we will be covering the following topics:

- What is NativeScript for Angular?
- Setting up your native mobile app
- Project Organization
- Architecture planning
- `@NgModule` decorator
- `@Injectable` decorator
- Breaking your app into modules

Mental preparation

Before diving right into coding, you can greatly enhance the development experience for your project by mapping out the various services and features your app needs. Doing so will help reduce code duplication, frame your data flow, and lead the way for rapid feature development in the future.

A service is a class that typically handles processing and/or provides data to your app. Your usage of these services does not need to know the specifics of where the data came from, just that it can ask the service for its purpose and it will happen.

The sketch exercise

A good exercise for this is to sketch out a rough idea of one of your app views. You may not know what it will look like yet and that's okay; this is purely an exercise to think about the user expectations as a first step to guiding your thought process into the various sections or modules you need to construct to meet those expectations. It will also help you think about the various states the app needs to manage.

Take, for example, the app we are going to build, **TNSStudio** (**Telerik NativeScript** (**TNS**)). We will dive into more detail of what our app is and what exactly it will do in `Chapter 2`, *Feature Modules*.

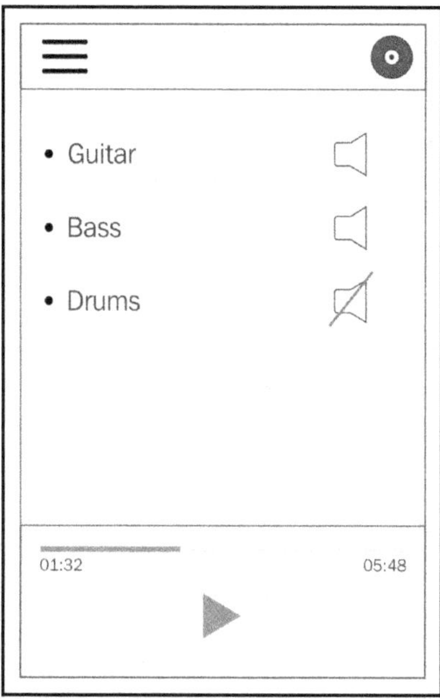

Starting from top to bottom, we can see a header with a menu button, a logo, and a record button. Then, we have a listing of user recorded tracks, each with a (re)record button and a solo or mute button.

From this one sketch, we may think about several services the app may need to provide:

- A Player Service
- A Recorder Service
- A Persistent Store service to remember which volume level settings the user sets for each track in the recording mix and/or if the user is authenticated

We can also gain some insight into the various states the app may need to manage:

- A listing of user recordings/tracks
- Whether the app is playing audio or not
- Whether the app is in the recording mode or not

Low-level thinking

It's also advantageous to provide some low-level services that provide a convenient API to access things, such as HTTP remote requests and/or logging. Doing so will allow you to create unique characteristics that you or your team like to work with when interacting with low-level APIs. For instance, maybe your backend API requires a unique header to be set in addition to a special authentication header for each request. Creating a low-level wrapper around an HTTP service will allow you to isolate those unique characteristics and provide a consistent API for your app to interact with, to guarantee all the API calls are enhanced with what they need in one place.

Additionally, your team may desire an ability to funnel all the logging code to a third-party log analyzer (for debugging or other performance-related metrics). Creating low-level wrappers with the lean code around some framework services will allow your app to adapt to these potential needs quickly.

Modularize with @NgModule

We can then think about breaking these services up into organizational units or modules.

Angular provides us with the `@NgModule` decorator, which will help us define what these modules look like and what they provide to our app. In an effort to keep our app's bootstrap/launch time as fast as possible, we can organize our modules in such a way to allow some service/features to be lazily loaded after our app has launched. Bootstrapping one module with a small subset of required code that our app needs to launch will help keep this launch phase to a minimum.

Our app's module breakdown

Here's how we will break down our app organization by module:

1. `CoreModule`: Low-level services, components, and utilities that provide a nice foundation layer. Things such as interacting with logging, dialogs, HTTP, and other various commonly used services.
2. `AnalyticsModule`**: Potentially, you could have a module that provides various services to handle analytics for your app.

3. `PlayerModule`*: Provides everything our app needs to play audio.
4. `RecorderModule`*: Provides everything our app needs to record audio.

*(*)*These are considered *Feature Modules.*
*(**)*We will omit this module from the example in this book but wanted to mention it here for context.

The module benefits

Using a similar organization provides several advantageous things for you and your team:

- **High degree of usability**: By designing a low-level `CoreModule`, you and your team have the opportunity to design how you like to work with commonly used services, in a unique way, across not only the app you are building now but more in the future. You can easily move `CoreModule` into a completely different app and gain all the same unique APIs you have designed for this app when working with low-level services.
- **Viewing your own app code as a 'Feature Module'**: Doing so will help you focus on just the unique abilities your app should provide outside of what the `CoreModule` provides as well as reduce the duplication of the code.
- **Encourages and enhances rapid development**: By confining commonly used functionality to our `CoreModule`, we relieve the burden of having to worry about those details in our feature modules. We can simply inject those services provided by our `CoreModule` and use those APIs and never repeat ourselves.
- **Maintainability**: In the future, if an underlying detail needs to change because of how your app needs to work with a low-level service, it need only be changed in one place (in the `CoreModule` service) versus having redundant code potentially spread across different sections of your app.
- **Performance**: Splitting your app into modules will allow you to load only the modules you need at startup, then later, lazily load other features on demand. Ultimately, this leads to a faster app startup time.

Considerations?

You may be thinking, why not just combine the player/recorder modules together into one module?

Answer: Our app is only going to allow recording when a registered user is authenticated. Therefore, it is beneficial to consider the potential of authenticated contexts and what features are only accessible to authenticated users (if any). This will allow us to further fine tune the loading performance of our app to what is needed when it's needed only.

Getting started

We are going to assume that you have NativeScript installed properly on your computer. If you do not, please follow the install instructions at `https://nativescript.org`. Once installed, we need to create our app framework using a shell prompt:

```
tns create TNSStudio --ng
```

The `tns` stands for Telerik NativeScript. It is the primary **command-line user interface** (**CLI**) tool you will use to create, build, deploy, and test any NativeScript app.

This command will create a new folder called `TNSStudio`. Inside is your primary project folder including everything required to build an app. It will contain everything relating to this project. After the project folder has been created, you need to do one more thing to have a fully runnable app. That's, adds the runtimes for Android and/or iOS:

```
cd TNSStudio
tns platform add ios
tns platform add android
```

If you are on a Macintosh, you can build for both iOS and Android. If you are running on a Linux or Windows device, Android is the only platform you can compile for on your local machine.

Create our module shells

Without writing the implementation of our services yet, we can define what our `CoreModule` will generally look like with `NgModule` by starting to define what it should provide:

Let's create `app/modules/core/core.module.ts`:

```
// angular
import { NgModule } from '@angular/core';
@NgModule({})
export class CoreModule { }
```

Injectable services

Now, let's create the boilerplate we need for our services. Note here that the injectable decorator is imported from Angular to declare that our service will be made available through Angular's **Dependency Injection** (**DI**) system, which allows these services to be injected into any class constructor that may need it. The DI system provides a nice way to guarantee that these services will be instantiated as singletons and shared across our app. It's also worth noting that we could alternatively provide these services on the component level if we didn't want them to be singletons and instead have unique instances created for certain branches of our component tree, which will make up our user interface. In this case, though, we want these created as singletons. We will be adding the following to our `CoreModule`:

- `LogService`: Service to funnel all our console logging through.
- `DatabaseService`: Service to handle any persistent data our app needs. For our app, we will implement the native mobile device's storage options, such as application settings, as a simple key/value store. However, you could implement more advanced storage options here, such as remote storage through Firebase for example.

Create `app/modules/core/services/log.service.ts`:

```
// angular
import { Injectable } from '@angular/core';
@Injectable()
export class LogService {
}
```

Also, create `app/modules/core/services/database.service.ts`:

```
// angular
import { Injectable } from '@angular/core';
@Injectable()
export class DatabaseService {
}
```

Consistency and standards

For consistency and to reduce the length of our imports as well as prepare for better scalability, let's also create an `index.ts` file in `app/modules/core/services`, which will export a `const` collection of our services as well as export these services (in an alphabetical order to keep things tidy):

```
import { DatabaseService } from './database.service';
import { LogService } from './log.service';

export const PROVIDERS: any[] = [
  DatabaseService,
  LogService
];

export * from './database.service';
export * from './log.service';
```

We will follow a similar pattern of the organization throughout the book.

Finalizing CoreModule

We can now modify our `CoreModule` to use what we have created. We will take this opportunity to also import the `NativeScriptModule` which our app will need to work with other NativeScript for Angular features which we will want accessible globally for our app. Since we know we will want those features, globally, we can also specify that they are exported so when we import and use our `CoreModule`, we won't need to worry about importing `NativeScriptModule` elsewhere. Here's what our `CoreModule` modifications should look like:

```
// nativescript
import { NativeScriptModule } from 'nativescript-
angular/nativescript.module';
// angular
import { NgModule } from '@angular/core';
```

```
// app
import { PROVIDERS } from './services';
@NgModule({
  imports: [
    NativeScriptModule
  ],
  providers: [
    ...PROVIDERS
  ],
  exports: [
    NativeScriptModule
  ]
})
export class CoreModule { }
```

We now have a good starting base for our `CoreModule`, the details of which we will implement in the following chapters.

Summary

We created a solid foundation for our app in this chapter. You learned how to think about your app's architecture in terms of modules. You also learned how to utilize Angular's `@NgModule` decorator to frame out these modules. And finally, we now have a great base architecture to work from to build our app on top of.

Now that you have some of the key concepts under your belt, we can now move onto the heart of our app, the feature modules. Let's dive into the main features of our app to continue constructing our service layers in `Chapter 2`, *Feature Modules*. We will soon be creating some views for our app and running the app on iOS and Android in `Chapter 3`, *Our First View via Component Building*.

2
Feature Modules

We are going to continue building the foundation of our app by scaffolding the core feature modules our app will need, the player and recorder. We will also want to keep in mind that the recording features will only be loaded and available when a user authenticates. Lastly, we will finish the implementation of our services from the `CoreModule` we created in `Chapter 1`, *Get Into Shape with @NgModule*.

In this chapter, we will be covering the following topics:

- Creating feature modules
- Separation of concerns with app features
- Setting up the `AppModule` to bootstrap efficiently, only loading upfront the feature modules we need for our first view
- Using the NativeScript `application-settings` module as our key/value store
- Providing the ability to control our app's debug logs at one spot
- Creating a new service that will use other services to demonstrate our scalable architecture

Player and recorder modules

Let's create the shell of our two main feature modules. Take note that we also add `NativeScriptModule` to the imports of both of the following modules:

1. `PlayerModule`: It will provide player-specific services and components that will be usable whether the user is authenticated or not.

   Let's create `app/modules/player/player.module.ts`:

   ```
   // nativescript
   import { NativeScriptModule } from 'nativescript-
   angular/nativescript.module';
   // angular
   import { NgModule, NO_ERRORS_SCHEMA } from '@angular/core';

   @NgModule({
     imports: [ NativeScriptModule ]
     schemas: [ NO_ERRORS_SCHEMA ]
   })
   export class PlayerModule { }
   ```

2. `RecorderModule`: This will provide recording-specific services and components that will only be loaded if the user is authenticated and enters the record mode for the first time.

   Let's create `app/modules/recorder/recorder.module.ts`:

   ```
   // nativescript
   import { NativeScriptModule } from 'nativescript-
   angular/nativescript.module';

   // angular
   import { NgModule, NO_ERRORS_SCHEMA } from '@angular/core';

   @NgModule({
     imports: [ NativeScriptModule ],
     schemas: [ NO_ERRORS_SCHEMA ]
   })
   export class RecorderModule { }
   ```

A shared model for our data

Before we go about creating our services, let's create an interface and model implementation for the core piece of data our app will be using. The `TrackModel` will represent a single track with the following:

- `filepath`: (to the local file)
- `name`: (for our view)
- `order`: Position (for the view listing of tracks)
- `volume`: We want our player to be able to mix different tracks together with different volume level settings
- `solo`: Whether we want to hear just this track in our mix

We will also add a convenient constructor to our model, which will take an object to initialize our model with.

Create `app/modules/core/models/track.model.ts`, since it will be shared across both our player and recorder:

```
export interface ITrack {
  filepath?: string;
  name?: string;
  order?: number;
  volume?: number;
  solo?: boolean;
}
export class TrackModel implements ITrack {
  public filepath: string;
  public name: string;
  public order: number;
  public volume: number = 1; // set default to full volume
  public solo: boolean;

  constructor(model?: any) {
    if (model) {
      for (let key in model) {
        this[key] = model[key];
      }
    }
  }
}
```

Scaffolding out the service APIs

Now, let's create the API our services will provide to our app. Starting with `PlayerService`, we could imagine the following API might be useful to manage tracks and control playback. Most of it should be fairly self-explanatory. We may refactor this later but this is a great start:

- `playing: boolean;`
- `tracks: Array<ITrack>;`
- `play(index: number): void;`
- `pause(index: number): void;`
- `addTrack(track: ITrack): void;`
- `removeTrack(track: ITrack): void;`
- `reorderTrack(track: ITrack, newIndex: number): void;`

Create `app/modules/player/services/player.service.ts` and stub out a few of the methods; some of them we could go ahead and implement:

```
// angular
import { Injectable } from '@angular/core';

// app
import { ITrack } from '../../core/models';
@Injectable()
export class PlayerService {
  public playing: boolean;
  public tracks: Array<ITrack>;
  constructor() {
    this.tracks = [];
  }

  public play(index: number): void {
    this.playing = true;
  }
  public pause(index: number): void {
    this.playing = false;
  }
  public addTrack(track: ITrack): void {
    this.tracks.push(track);
  }
  public removeTrack(track: ITrack): void {
    let index = this.getTrackIndex(track);
    if (index > -1) {
      this.tracks.splice(index, 1);
```

```
      }
    }
    public reorderTrack(track: ITrack, newIndex: number) {
      let index = this.getTrackIndex(track);
      if (index > -1) {
        this.tracks.splice(newIndex, 0, this.tracks.splice(index, 1)[0]);
      }
    }
    private getTrackIndex(track: ITrack): number {
      let index = -1;
      for (let i = 0; i < this.tracks.length; i++) {
        if (this.tracks[i].filepath === track.filepath) {
          index = i;
          break;
        }
      }
      return index;
    }
  }
```

Now, let's apply our standard by exporting this service for our module.

Create `app/modules/player/services/index.ts`:

```
import { PlayerService } from './player.service';

export const PROVIDERS: any[] = [
  PlayerService
];

export * from './player.service';
```

Lastly, modify our `PlayerModule` to specify the correct providers so our final module should look like the following:

```
// nativescript
import { NativeScriptModule } from 'nativescript-
angular/nativescript.module';

// angular
import { NgModule, NO_ERRORS_SCHEMA } from '@angular/core';

// app
import { PROVIDERS } from './services';

@NgModule({
  imports: [ NativeScriptModule ],
  providers: [ ...PROVIDERS ],
```

```
  schemas: [ NO_ERRORS_SCHEMA ]
})
export class PlayerModule { }
```

Next, we can design `RecorderService` to provide a simple recording API.

Create `app/modules/recorder/services/recorder.service.ts`:

- `record(): void`
- `stop(): void`

```
// angular
import { Injectable } from '@angular/core';
@Injectable()
export class RecorderService {
  public record(): void { }
  public stop(): void { }
}
```

Now, apply our standard by exporting this service for our module.

Create `app/modules/recorder/services/index.ts`:

```
import { RecorderService } from './recorder.service';

export const PROVIDERS: any[] = [
  RecorderService
];

export * from './recorder.service';
```

Lastly, modify our `RecorderModule` to specify the correct providers so our final module should look like the following:

```
// nativescript
import { NativeScriptModule } from 'nativescript-
angular/nativescript.module';

// angular
import { NgModule, NO_ERRORS_SCHEMA } from '@angular/core';

// app
import { PROVIDERS } from './services';

@NgModule({
  imports: [ NativeScriptModule ],
  providers: [ ...PROVIDERS ],
```

```
  schemas: [ NO_ERRORS_SCHEMA ]
})
export class RecorderModule { }
```

With our two primary feature modules scaffolded and ready to go, let's revisit the two low-level services we created in `Chapter 1`, *Get Into Shape with @NgModule*, and provide implementations.

Implementing LogService

Logging is an important ally that you want during the development lifecycle of your app as well as in production. It can help you debug as well as gain important insights into how your app is used. Having a single pathway to run all logging through also provides an opportunity to reroute all the app logs somewhere else with the flip of a switch. For example, you could use a third-party debug tracking service, such as TrackJS (`https://trackjs.com`), via Segment (`https://segment.com`). You will want to run a lot of important aspects of your app through logging and it serves as a great place to have a lot of control and flexibility.

Let's open `app/modules/core/services/log.service.ts` and get to work. Let's start by defining a static boolean that will serve as a simple flag we can toggle in our `AppModule` to enable/disable. Let's also add a few helpful methods:

```
import { Injectable } from '@angular/core';

@Injectable()
export class LogService {

 public static ENABLE: boolean = true;

 public debug(msg: any, ...formatParams: any[]) {
   if (LogService.ENABLE) {
     console.log(msg, formatParams);
   }
 }

 public error(msg: any, ...formatParams: any[]) {
   if (LogService.ENABLE) {
     console.error(msg, formatParams);
   }
 }

 public inspect(obj: any) {
   if (LogService.ENABLE) {
```

```
        console.log(obj);
        console.log('typeof: ', typeof obj);
        if (obj) {
          console.log('constructor: ', obj.constructor.name);
          for (let key in obj) {
            console.log(`${key}: `, obj[key]);
          }
        }
      }
    }
  }
```

- `debug`: This will serve as our most commonly used output API for logging.
- `error`: When we know a certain condition is an error, this will help identify those spots in our log.
- `inspect`: There are times when viewing an object can help find a bug or help us understand the state of our app at any given moment.

With our `LogService` implemented, we will now use it throughout our app and the rest of this book instead of using the console directly.

Implementing DatabaseService

Our `DatabaseService` needs to provide several things:

- A persistent store to save and retrieve any data our app needs.
- It should allow any type of data to be stored; however, we will specifically want it to handle JSON serialization.
- Static keys of all the data we will want to store.
- A static reference to a saved user? Well, yes it could. However, this brings up a point that we will address in a moment.

Regarding the first item, we can use NativeScript's `application-settings` module. Under the hood, this module provides a consistent API to work with two native mobile APIs:

- **iOS**: `NSUserDefaults`: `https://developer.apple.com/reference/foundation/userdefaults`
- **Android**: `SharedPreferences`: `https://developer.android.com/reference/android/content/SharedPreferences.html`

Regarding serializing JSON data, the `application-settings` module provides a `setString` and `getString` method, which will allow us to use it in conjunction with `JSON.stringify` and `JSON.parse`.

Using string values throughout your codebase in several different spots to refer to the same key that should remain constant can become error prone. Because of this, we will keep a typed (for type safety) static hash of valid keys that our app will use. We may only know one at this point in time (authenticated user as `'current-user'`) but creating this will provide a single spot to scale these out over time.

Four? We will address four in a moment.

Open `app/modules/core/services/database.service.ts` and modify it to provide a similar API to the web's `localStorage` API for simplicity:

```
// angular
import { Injectable } from '@angular/core';

// nativescript
import * as appSettings from 'application-settings';

interface IKeys {
  currentUser: string;
}
@Injectable()
export class DatabaseService {

  public static KEYS: IKeys = {
    currentUser: 'current-user'
  };

  public setItem(key: string, value: any): void {
    appSettings.setString(key, JSON.stringify(value));
  }

  public getItem(key: string): any {
    let item = appSettings.getString(key);
    if (item) {
      return JSON.parse(item);
    }
    return item;
  }

  public removeItem(key: string): void {
    appSettings.remove(key);
  }
}
```

This service now provides a way to store an object via `setItem`, which ensures the object is properly stored as a string via `JSON.stringify`. It also provides a way to retrieve values via `getItem`, which also handles the serialization back to an object for us via `JSON.parse`. We also have the `remove` API to simply remove values from our persisted store. Lastly, we have a nice static reference to all the valid keys that our persistent store will keep track of.

Now, what about that static reference to the saved user?

We want to be able to easily access our authenticated user from anywhere in the app. We could provide a static reference in our `DatabaseService` for simplicity, but our aim here is to have a clear separation of concerns. Since we know we will want the ability to show a modal asking the user to register and unlock those recording features, a new service to manage this makes sense. Since we have designed scalable architecture, we can easily add another service into the mix, so let's do that now!

Create AuthService to help handle the authenticated state of our app

One important consideration for our `AuthService` is to understand that certain components in our app may benefit from getting notified when the authenticated state changes. This is a perfect use case to utilize RxJS. RxJS is a very powerful library that is used to simplify dealing with changing data and events using observables. An observable is a data type that you can use not only to listen to events, but filter, map, reduce, and run sequences of code against anytime something occurs. By using observables, we can simplify our asynchronous development dramatically. We will use a specific type of observable called the `BehaviorSubject` to emit changes that our components could subscribe to.

Create `app/modules/core/services/auth.service.ts` and add the following:

```
// angular
import { Injectable } from '@angular/core';

// lib
import { BehaviorSubject } from 'rxjs/BehaviorSubject';

// app
import { DatabaseService } from './database.service';
import { LogService } from './log.service';

@Injectable()
export class AuthService {

  // access our current user from anywhere
```

```
public static CURRENT_USER: any;

// subscribe to authenticated state changes
public authenticated$: BehaviorSubject<boolean> =
  new BehaviorSubject(false);

constructor(
  private databaseService: DatabaseService,
  private logService: LogService
) {
  this._init();
}

private _init() {
  AuthService.CURRENT_USER = this.databaseService
    .getItem(DatabaseService.KEYS.currentUser);
  this.logService.debug(`Current user: `,
    AuthService.CURRENT_USER);
  this._notifyState(!!AuthService.CURRENT_USER);
}

private _notifyState(auth: boolean) {
  this.authenticated$.next(auth);
}
}
```

We have a few interesting things going on here. We are putting two other services we designed to work right away, `LogService` and `DatabaseService`. They are helping us check whether a user was saved/authenticated as well as log that result.

We are also calling on a `private _init` method when our service gets constructed via Angular's dependency injection system. This allows us to immediately check whether an authenticated user exists in our persistent store. Then, we call a private reusable method `_notifyState`, which will emit `true` or `false` on our `authenticated$` observable. This will provide a nice way for other components to easily get notified when the auth state changes by subscribing to this observable. We have made `_notifyState` reusable because our login and register methods (to be implemented in the future) will be able to use it when the results are returned from modals we may display in the UI.

We can now easily add `AuthService` to our `PROVIDERS` and we don't need to do anything else to ensure it's added to our `CoreModule` because our `PROVIDERS` are already added to the `CoreModule`.

All we need to do is modify `app/modules/core/services/index.ts` and add our service:

```
import { AuthService } from './auth.service';
import { DatabaseService } from './database.service';
import { LogService } from './log.service';

export const PROVIDERS: any[] = [
 AuthService,
 DatabaseService,
 LogService
];

export * from './auth.service';
export * from './database.service';
export * from './log.service';
```

WAIT! There is one important thing we want to do to ensure our AuthService initializes!

Angular's dependency injection system will only instantiate a service that is injected somewhere. Although we have all our services specified as providers in our `CoreModule`, they will not actually be constructed until they are injected somewhere!

Open `app/app.component.ts` and replace its contents with this:

```
// angular
import { Component } from '@angular/core';

// app
import { AuthService } from './modules/core/services';

@Component({
 selector: 'my-app',
 templateUrl: 'app.component.html',
})
export class AppComponent {

 constructor(private authService: AuthService) { }

}
```

We inject our `AuthService` by specifying it as an argument to our component's constructor. This will cause Angular to construct our service. All subsequent injects throughout our code will all receive the same singleton.

Prepare to bootstrap the AppModule

We now have a good setup for our feature modules and it's time to bring them all together in our root `AppModule` responsible for bootstrapping our app.

Bootstrap only what is needed for your initial view. Lazy load the rest.

It's important to keep the bootstrap of our app as fast as possible. To achieve that, we only want to bootstrap the app with the main features needed for our initial view and lazy load the rest when needed. We know we want our low-level services to be available and ready to use anywhere in the app, so we will definitely want `CoreModule` upfront.

Our initial view from our sketch is going to start with the player and 2-3 tracks on the list, so the user can immediately playback a mix of pre-recorded tracks we will ship with the app for demonstration purposes. For this reason, we will specify the `PlayerModule` to load upfront when our app bootstraps, since it will be a primary feature we want to immediately engage with.

We will set up a routing configuration, which will lazy load our `RecorderModule` when the user taps the record button at the top right of our initial view to begin a recording session.

With this in mind, we can set up our `AppModule` located at `app/app.module.ts`, as follows:

```
// angular
import { NgModule } from '@angular/core';

// app
import { AppComponent } from './app.component';
import { CoreModule } from './modules/core/core.module';
import { PlayerModule } from './modules/player/player.module';

@NgModule({
  imports: [
    CoreModule,
    PlayerModule
  ],
  declarations: [AppComponent],
  bootstrap: [AppComponent]
})
export class AppModule { }
```

Summary

Throughout, we have been working hard creating a solid foundation to build our app on. We created a `CoreModule` to provide some low-level services, such as logging, and a persistent store and designed the module to easily scale in more services as needed . Plus, this module is portable and can be dropped into other projects with your own company's special sauce intact.

In typical app development, you may want to run your app on the iOS and/or Android simulator along the way, during this process to double-check some of your design/architecture choices and that would be advisable! We just haven't done that yet, since we have an app pre-planned here and want you to stay focused on the choices we are making and why.

We also created the two primary feature modules that our app needs for its core competency, `PlayerModule` and `RecorderModule`. The player will be pre-setup with 2-3 recorded tracks loaded and ready to play right upon launch, so we will be bootstrapping our app with the `PlayerModule` features.

We will provide a simple way to allow a user to register an account, which will allow them to record their own tracks to throw in the mix. Once they are logged in, they will be able to enter the record mode via a route, which will lazily load the `RecorderModule`.

In the next chapter, we will create our first view, configure our routes, and finally, get our first glimpse at our app.

3

Our First View via Component Building

We've been working hard at framing the base of our app in `Chapter 2`, *Feature Modules*, and now it's time to finally get a glimpse of what we're working with. This is all about getting that first view from our sketch to the mobile device screen.

Building views with NativeScript for Angular is not much different than view building for the web. We will use Angular's Component decorator to build various components our UI needs to achieve the desired usability we're after. Instead of using the HTML markup, we will be using NativeScript XML, which is an extremely powerful, yet simple and concise, abstraction of all native view components on both iOS and Android.

We won't be covering all the benefits and types of components you have access to here; but to learn more, we recommend any of the following books:

- `https://www.packtpub.com/web-development/getting-started-nativescript`
- `https://www.manning.com/books/nativescript-in-action`

In this chapter, we will be covering the following topics:

- Using Component decorator to compose our views
- Creating reusable components
- Creating a custom view filter using Pipe
- Running the app on the iOS and Android simulators

Our first view via component building

If we look at our sketch from `Chapter 1`, *Get into Shape with @NgModule*, we can see a header at the top of the app, which will contain our app title with the record button to the right. We also see a listing of tracks with some player controls at the bottom. We can break these key elements of our UI design into essentially three primary components. One component is already provided by the NativeScript framework, the ActionBar, which we will use to represent the top header.

NativeScript provides many rich view components to build our UI. The markup is not HTML but rather XML with an `.html` extension, which may seem unusual. The reason the `.html` extension is used for XML view templates with NativeScript for Angular is that the custom renderer (`https://github.com/NativeScript/nativescript-angular`) uses a DOM adapter to parse the view template. Each NativeScript XML component represents true native view widgets on each respective platform.

For the other two primary components, we will use Angular's Component decorator. It's important at this phase of the app development cycle to think about encapsulated pieces of UI functionality. We will encapsulate our track listing as a component and the player controls as another component. In this exercise, we will use an outside-in approach to building our UI from an abstract viewpoint down to the implementation details of each component.

To begin, let's focus on the root component in our Angular app because it will define the basic layout of our first view. Open `app/app.component.html`, clear its contents, and replace with the following to rough out the initial UI concept from our sketch:

```
<ActionBar title="TNSStudio">
</ActionBar>
<GridLayout rows="*, 100" columns="*">
  <track-list row="0" col="0"></track-list>
  <player-controls row="1" col="0"></player-controls>
</GridLayout>
```

We are expressing our view with `ActionBar` and the primary layout container for the main view, `GridLayout`. With NativeScript, it's important that each view starts with a layout container as the root node (outside of any `ActionBar` or `ScrollView`), much like `div` tags that are used with HTML markup. At the time of this writing, there are six layout containers provided by NativeScript: `StackLayout`, `GridLayout`, `FlexboxLayout`, `AbsoluteLayout`, `DockLayout`, and `WrapLayout`. For our layout here, `GridLayout` will work well.

All about the GridLayout

The GridLayout is one of the three most used layouts you will use in your NativeScript application (the others are FlexboxLayout and StackLayout). This is the layout that allows you to build complex layouts easily. To use the GridLayout is very much like the enhanced table in HTML. You are basically going to want to take your screen area and divide your screen into the pieces you need. It will allow you to tell the columns (or rows) to be a percentage of the remaining width (and height) of the screen. The grid supports three types of values; **absolute size**, a percentage of **remaining space**, and **used space**.

For **absolute size**, you just type in the number. For example, `100` means it will use 100 dp of space.

Another name for **dp** is **dip**. They are the same. A device-independent pixel (also density-independent pixel, DIP, or DP) is a physical unit of measurement based on a coordinate system held by a computer and represents an abstraction of a pixel for use by an application that an underlying system then converts to physical pixels.

If you take the smallest iOS device supported, it has a screen width of 320dp. For other devices, such as tablets, some have a width of 1024 dp. So, 100 dp would be almost one third of an iOS phone, where it is one tenth of the screen on a tablet. So, this is something you need to think about when using fixed absolute values. It is typically better to use the used space over a fixed value, unless you are needing to constrain the column to a specific size.

To use **remaining space-**based values , that is, `*`, the `*` tells it to use the rest of the remaining space. If the columns (or rows) is set to `*, *`, then space will be divided into two equal remainders of space. Likewise, `rows="*,*,*,*,*"` will specify five equal sized rows. You can also specify things, such as `columns="2*,3*,*"`, and you will get three columns; the first column will be two sixth of the screen, the second column will be three sixth of the screen, and the final column will be one sixth of the screen (that is, 2+3+1 = 6). This allows you great flexibility in how to use the remainder of the space.

The third type of sizing is **space used**. So what happens is the content inside the grid is measured and then the column is assigned the size that is the max used in that column (or row). This is very useful when you have a grid where you have data but you aren't sure of the size or you don't really care; you just want it to look good. So, this is the auto keyword. I might have `columns="auto,auto,*,auto"`. This means columns 1,2, and 4 will all be automatically sized based on the content inside those columns; and column 3 will use whatever space is left over. This is very useful for laying out the entire screen or parts of the screen where you are looking for a certain look.

The final reason why the GridLayout is one of the best layouts is that when you assign items to the GridLayout, you can actually assign multiple items to the same rows and/or columns and you can use row or column spans to allow items to use more than one row and/or column.

To assign an object, you just assign it via `row="0"` and/or `col="0"` (keep in mind these are index-based positions). You can also use `rowSpan` and `colSpan` to make an element span multiple rows and/or columns. Overall, the GridLayout is the most versatile layout and allows you to easily create almost any layout you will need in your app.

Back to our layout

Inside the grid, we have declared a `track-list` component to represent our track listing, which will flex vertically, taking up all the vertical space and leaving only a height of 100 for `player-controls`. We indicate `track-list` as `row="0" col="0"`, since rows and columns are index-based. The flexible (remainder) vertical height is defined via the GridLayout's `*` in the rows attribute. The bottom section of the grid (row 1) will represent the player controls, allowing users to play/pause the mix and shuttle the playback position.

Now that we have the app's primary view defined in a rather abstract way, let's dive into the two custom components we need to build, `track-list` and `player-controls`.

Building TrackList component

The track list should be a listing of all the recorded tracks. Each row in the list should provide a single record button to re-record in addition to a name label for displaying the title provided by the user. It should also provide a switch to allow the user to solo just that particular track.

We can inject `PlayerService` and declare it `public` to allow us to bind directly to the service's tracks collection.

We can also mock out some of our bindings to get things rolling like the `record` action. For now, let's just allow a track to be passed in and let's print out an inspection of that track via `LogService`.

Let's start by creating `app/modules/player/components/track-list/ track-list.component.ts` (with a matching `.html` template):

```
// angular
import { Component, Input } from '@angular/core';

// app
import { ITrack } from '../../../core/models';
import { LogService } from '../../../core/services';
import { PlayerService } from '../../services/player.service';

@Component({
 moduleId: module.id,
 selector: 'track-list',
 templateUrl: 'track-list.component.html'
})
export class TrackListComponent {

 constructor(
   private logService: LogService,
   public playerService: PlayerService
 ) { }

 public record(track: ITrack) {
   this.logService.inspect(track);
 }
}
```

For the view template `track-list.component.html`, we are going to employ the powerful `ListView` component. This widget represents the native UITableView (`https://developer.apple.com/reference/uikit/uitableview`) on iOS and the native ListView (`https://developer.android.com/guide/topics/ui/layout/listview.html`) on Android, offering 60 fps virtual scrolling with reused rows. Its performance is unparalleled on mobile devices:

```
<ListView [items]="playerService.tracks">
  <ng-template let-track="item">
    <GridLayout rows="auto" columns="75,*,100">
      <Button text="Record" (tap)="record(track)"
          row="0" col="0"></Button>
      <Label [text]="track.name" row="0" col="1"></Label>
      <Switch [checked]="track.solo" row="0" col="2">
      </Switch>
    </GridLayout>
  </ng-template>
</ListView>
```

There's a lot going on with this view template, so let's inspect it a bit.

Since we made `playerService` `public` upon injection into our Component's constructor, we can bind directly to its tracks via the `ListView` items' attribute using standard Angular binding syntax expressed as `[items]`. This will be the collection our list will iterate on.

The `template` node inside allows us to encapsulate how each row of our list will be laid out. It also allows us to declare a variable name (`let-track`) for use as our iterator reference.

We start with a GridLayout, since each row will contain a **Record** button (to allow a track to be re-recorded), to which we will assign a width of 75. This button will be bound to the `tap` event, which will activate a recording session if the user is authenticated.

Then, we will have a Label to display a user-provided name for the track, which we will assign `*` to ensure it expands to fill the horizontal space in between our left-hand and right-hand columns. We use the text attribute to bind to `track.name`.

Lastly, we will use `switch` to allow the user to toggle soloing the track in the mix. This provides the `checked` attribute to allow us to bind our `track.solo` property to.

Building a dialog wrapper service to prompt the user

If you recall from `Chapter 1`, *Get Into Shape with @NgModule,* recording is a feature that should only be available to authenticated users. Therefore, we will want to prompt the user with a login dialog when they tap the **Record** button on each track. If they are already logged in, we will want to prompt them to confirm if they want to re-record the track for good usability.

We could handle this dialog directly in the Component by importing a NativeScript dialog service that provides a consistent API across both platforms. The `ui/dialogs` module from the NativeScript framework (`https://docs.nativescript.org/ui/dialogs`) is a very convenient service, allowing you to create native alerts, confirms, prompts, actions, and basic login dialogs. However, we may want to provide custom native dialog implementations on both iOS and Android down the road for an even nicer UX experience. There are several plugins that provide very elegant native dialogs, for example, `https://github.com/NathanWalker/nativescript-fancyalert`.

To prepare for this enriched user experience, let's build a quick Angular service that we can inject and use everywhere, which will allow us to easily implement these niceties down the road.

Since this should be considered a `core` service to our app, let's create `app/modules/core/services/dialog.service.ts`:

```
// angular
import { Injectable } from '@angular/core';

// nativescript
import * as dialogs from 'ui/dialogs';

@Injectable()
export class DialogService {

  public alert(msg: string) {
    return dialogs.alert(msg);
  }

  public confirm(msg: string) {
    return dialogs.confirm(msg);
  }

  public prompt(msg: string, defaultText?: string) {
    return dialogs.prompt(msg, defaultText);
  }

  public login(msg: string, userName?: string, password?: string) {
    return dialogs.login(msg, userName, password);
  }

  public action(msg: string, cancelButtonText?: string,
    actions?: string[]) {
    return dialogs.action(msg, cancelButtonText, actions);
  }
}
```

At first glance, this may appear incredibly wasteful! Why create a wrapper that provides the exact same API as a service that already exists from the NativeScript framework?

Yes, indeed, at this stage, it appears that way. However, we are preparing for greatness in flexibility and power with how we will handle these dialogs in the future. Stay tuned for a potential bonus chapter of material covering this fun and unique polish to the integration.

The last thing we need to do before we move on to use this service is to ensure it is added to our core service `PROVIDERS` collection. This will make sure Angular's DI system knows our new service is a valid token available for injection.

Open app/modules/core/services/index.ts and modify as follows:

```
import { AuthService } from './auth.service';
import { DatabaseService } from './database.service';
import { DialogService } from './dialog.service';
import { LogService } from './log.service';

export const PROVIDERS: any[] = [
 AuthService,
 DatabaseService,
 DialogService,
 LogService
];

export * from './auth.service';
export * from './database.service';
export * from './dialog.service';
export * from './log.service';
```

We are now ready to inject and use our new service.

Integrating DialogService into our component

Let's open up track-list.component.ts and inject DialogService for use in our record method. We will also need to determine if the user is logged in to conditionally display a login dialog or confirm prompt, so let's also inject AuthService:

```
// angular
import { Component, Input } from '@angular/core';

// app
import { ITrack } from '../../../core/models';
import { AuthService, LogService, DialogService } from
'../../../core/services';
import { PlayerService } from '../../services/player.service';

@Component({
  moduleId: module.id,
  selector: 'track-list',
  templateUrl: 'track-list.component.html'
})
export class TrackListComponent {

 constructor(
   private authService: AuthService,
   private logService: LogService,
```

```
    private dialogService: DialogService,
    public playerService: PlayerService
  ) { }

  public record(track: ITrack, usernameAttempt?: string) {
    if (AuthService.CURRENT_USER) {
      this.dialogService.confirm(
        'Are you sure you want to re-record this track?'
      ).then((ok) => {
        if (ok) this._navToRecord(track);
      });
    } else {
      this.authService.promptLogin(
        'Provide an email and password to record.',
        usernameAttempt
      ).then(
        this._navToRecord.bind(this, track),
        (usernameAttempt) => {
          // initiate sequence again
          this.record(track, usernameAttempt);
        }
      );
    }
  }

  private _navToRecord(track: ITrack) {
    // TODO: navigate to record screen
    this.logService.debug('yes, re-record', track);
  }
}
```

The record method now first checks to see whether a user is authenticated via the static `AuthService.CURRENT_USER` reference, which is set when `AuthService` is first constructed via Angular's dependency injection upon app launch (see `Chapter 2`, *Feature Modules*).

If a user is authenticated, we present a confirmation dialog to ensure the action was intentional.

If the user is not authenticated, we want to prompt the user to log in. To reduce the overload for this book, we will assume the user is already registered via a backend API, so we won't be asking the user to register.

We need to implement the `promptLogin` method in `AuthService` to persist the user's login credentials, so every time they return to the app, it will automatically log them in. The record method now provides an extra optional argument `usernameAttempt`, which will be useful to repopulate the username field of the login prompt when reinitiating the login sequence after a user input validation error. We won't do a thorough validation of user input here, but we can at least do a lightweight check for a valid email.

In your own app, you should probably do more user input validation.

To maintain a clean separation of concerns, open `app/modules/core/services/auth.service.ts` to implement `promptLogin`. Here's the entire service with the modifications:

```
// angular
import { Injectable } from '@angular/core';

// lib
import { BehaviorSubject } from 'rxjs/BehaviorSubject';

// app
import { DatabaseService } from './database.service';
import { DialogService } from './dialog.service';
import { LogService } from './log.service';

@Injectable()
export class AuthService {

 // access our current user from anywhere
 public static CURRENT_USER: any;

 // subscribe to authenticated state changes
 public authenticated$: BehaviorSubject<boolean> =
   new BehaviorSubject(false);

 constructor(
 private databaseService: DatabaseService,
 private dialogService: DialogService,
 private logService: LogService
 ) {
   this._init();
 }

 public promptLogin(msg: string, username: string = '')
```

```
  : Promise<any> {
  return new Promise((resolve, reject) => {
    this.dialogService.login(msg, username, '')
      .then((input) => {
        if (input.result) { // result = false when canceled
          if (input.userName &&
              input.userName.indexOf('@') > -1) {
              if (input.password) {
                // persist user credentials
                this._saveUser(
                  input.userName, input.password
                );
                resolve();
              } else {
                this.dialogService.alert(
                  'You must provide a password.'
                ).then(reject.bind(this, input.userName));
              }
          } else {
            // reject, passing userName back
            this.dialogService.alert(
              'You must provide a valid email address.'
            ).then(reject.bind(this, input.userName));
          }
        }
      });
    });
}

private _saveUser(username: string, password: string) {
  AuthService.CURRENT_USER = { username, password };
  this.databaseService.setItem(
    DatabaseService.KEYS.currentUser,
    AuthService.CURRENT_USER
  );
  this._notifyState(true);
}

 private _init() {
   AuthService.CURRENT_USER =
     this.databaseService
     .getItem(DatabaseService.KEYS.currentUser);
   this.logService.debug(
     `Current user: `, AuthService.CURRENT_USER
   );
   this._notifyState(!!AuthService.CURRENT_USER);
 }
```

```
  private _notifyState(auth: boolean) {
    this.authenticated$.next(auth);
  }
}
```

We use the `dialogService.login` method to open a native login dialog, allowing the user to input a username and password. Once they choose ok, we do minimal validation of the input and, if successful, proceed to persist the username and password via `DatabaseService`. Otherwise, we simply alert the user of an error and reject our promise, passing along the username that was entered. This allows us to help the user out by redisplaying the login dialog with the failed username they entered, so they can more easily make corrections.

With these service level details complete, the `track-list` component is looking pretty good. However, there is one additional step we should take while we are working on this. If you recall, our TrackModel contains an order property that will help the user order the tracks in any way they'd like for convenience.

Creating an Angular Pipe - OrderBy

Angular provides the Pipe decorator for ease in creating view filters. Let's start by showing how we will use this in the view. You can see that it appears very similar to a command-line pipe used in Unix shell scripts; hence, it's named: `Pipe`:

```
<ListView [items]="playerService.tracks | orderBy: 'order'">
```

This will take the `playerService.tracks` collection and ensure it is ordered via the `order` property of each `TrackModel` for the view display.

Since we may want to use this anywhere in our app views, let's add this pipe as part of `CoreModule`. Create `app/modules/core/pipes/order-by.pipe.ts` and here is how we will implement `OrderByPipe`:

```
import { Pipe } from '@angular/core';

@Pipe({
 name: 'orderBy'
})
export class OrderByPipe {

 // Comparator method
 static comparator(a: any, b: any): number {
   if (a === null || typeof a === 'undefined') a = 0;
   if (b === null || typeof b === 'undefined') b = 0;
```

```
    if ((isNaN(parseFloat(a)) || !isFinite(a)) ||
        (isNaN(parseFloat(b)) || !isFinite(b))) {
       // lowercase strings
       if (a.toLowerCase() < b.toLowerCase()) return -1;
       if (a.toLowerCase() > b.toLowerCase()) return 1;
    } else {
      // ensure number values
      if (parseFloat(a) < parseFloat(b)) return -1;
      if (parseFloat(a) > parseFloat(b)) return 1;
    }

    return 0; // values are equal
  }

  // Actual value transformation
  transform(value: Array<any>, property: string): any {
    return value.sort(function (a: any, b: any) {
      let aValue = a[property];
      let bValue = b[property];
      let comparison = OrderByPipe
                       .comparator(aValue, bValue);
      return comparison;
    });
  }
}
```

We won't go into too much detail with what is going on here, since this is pretty typical in JavaScript to order a collection. To finish this off, ensure `app/modules/core/pipes/index.ts` follows our standard convention:

```
import { OrderByPipe } from './order-by.pipe';

export const PIPES: any[] = [
 OrderByPipe
];
```

Lastly, import the preceding collection for use with `app/modules/core/core.module.ts`. Here is the full file with all the modifications:

```
// nativescript
import { NativeScriptModule } from 'nativescript-
angular/nativescript.module';

// angular
import { NgModule } from '@angular/core';

// app
import { PIPES } from './pipes';
```

```
import { PROVIDERS } from './services';

@NgModule({
 imports: [
   NativeScriptModule
 ],
 declarations: [
   ...PIPES
 ],
 providers: [
   ...PROVIDERS
 ],
 exports: [
   NativeScriptModule,
   ...PIPES
 ]
})
export class CoreModule { }
```

Since pipes are view level implementations, we ensure they are added as part of the `exports` collection to allow other modules to use them.

Now, if we were to run our app at this point, you would notice that our `OrderBy` pipe used on our `track-list.component.html` view template would *NOT* work!

Angular modules compile in isolation of one another.

This is a critical point to understand: Angular compiles `PlayerModule` that declares `TrackListComponent` unto itself in an isolated sense. Since we declared `OrderByPipe` as part of `CoreModule` and `PlayerModule` has no dependency (at the moment) on `CoreModule`, the `TrackListComponent` gets compiled with no awareness of `OrderByPipe`! You would end up seeing this error generated in the console:

```
CONSOLE ERROR file:///app/tns_modules/tns-core-
modules/trace/trace.js:160:30: ns-renderer: ERROR BOOTSTRAPPING ANGULAR
CONSOLE ERROR file:///app/tns_modules/tns-core-
modules/trace/trace.js:160:30: ns-renderer: Template parse errors:
 The pipe 'orderBy' could not be found ("
 </ListView>-->

 <ListView [ERROR ->][items]="playerService.tracks | orderBy: 'order'">
   <ng-template let-track="item">
     <GridLayout rows"): TrackListComponent@10:10
```

To remedy this, we want to make sure `PlayerModule` is aware of view-related declarations (such as pipes or other components) from `CoreModule` by ensuring `CoreModule` is added as part of the `imports` collection on `PlayerModule`. This also provides us with one additional convenience. If you notice, `CoreModule` specifies `NativeScriptModule` as an export, which means any module that imports `CoreModule` will inherently get `NativeScriptModule` along with it. Here are the final modifications to `PlayerModule` to allow everything to work together:

```
// angular
import { NgModule } from '@angular/core';

// app
import { CoreModule } from '../core/core.module';
import { COMPONENTS } from './components';
import { PROVIDERS } from './services';

@NgModule({
 imports: [
   CoreModule
 ],
 providers: [...PROVIDERS],
 declarations: [...COMPONENTS],
 exports: [...COMPONENTS]
})
export class PlayerModule { }
```

We can now move on to the `player-controls` component.

Building PlayerControls component

Our player controls should contain a play/pause toggle button for the entire mix. It should also present a slider control to allow us to skip ahead and rewind our playback.

Let's create `app/modules/player/components/player-controls/player-controls.component.html` (with a matching `.ts`):

```
<GridLayout rows="100" columns="75,*" row="1" col="0">
  <Button [text]="playStatus" (tap)="togglePlay()" row="0"
col="0"></Button>
  <Slider minValue="0" [maxValue]="duration"
          [value]="currentTime" row="0" col="1"></Slider>
</GridLayout>
```

We start with a single row `GridLayout` with an explicit 100 height. Then, the first column will be constrained to 75 wide to accommodate our play/pause toggle button. Then, the second column will take up the rest of the horizontal space, indicated with `*` with the `Slider` component. This component is provided by the NativeScript framework and allows us to bind the `maxValue` attribute to the total duration of our mix as well as a value to `currentTime` of the playback.

Then, for `player-controls.component.ts`:

```
// angular
import { Component, Input } from '@angular/core';

// app
import { ITrack } from '../../../core/models';
import { LogService } from '../../../core/services';
import { PlayerService } from '../../services';

@Component({
 moduleId: module.id,
 selector: 'player-controls',
 templateUrl: 'player-controls.component.html'
})
export class PlayerControlsComponent {

 public currentTime: number = 0;
 public duration: number = 0;
 public playStatus: string = 'Play';

 constructor(
   private logService: LogService,
   private playerService: PlayerService
 ) { }

 public togglePlay() {
   let playing = !this.playerService.playing;
   this.playerService.playing = playing;
   this.playStatus = playing ? 'Stop' : 'Play';
 }

}
```

For now, we have placed `currentTime` and `duration` directly on the component, however, we will refactor those into `PlayerService` later. Eventually, all of the state related to our player will come from `PlayerService` when we implement plugins to handle our audio in subsequent chapters. The `togglePlay` method also just stubs out some general behavior, toggling the text of our button to **Play** or **Stop**.

Quick preview

At this point, we will take a quick look at what we have built so far. Currently, our player service returns an empty list of tracks. To see the results, we should add some dummy data to it. For example, in `PlayerService`, we could add:

```
constructor() {
  this.tracks = [
    {name: "Guitar"},
    {name: "Vocals"},
  ];
}
```

Don't be surprised if it's not pretty; we'll cover that in the next chapter. We also won't cover all the runtime commands available to us yet; we'll cover that thoroughly in `Chapter 6`, *Running the app on iOS and Android.*

Preview on iOS

You will have to be on a Mac with XCode installed to preview the iOS app:

```
tns run ios --emulator
```

This should launch the iOS Simulator and you should see the following screenshot:

Preview on Android

You will have to have the AndroidSDKk and tools installed to preview on an Android emulator:

```
tns run android --emulator
```

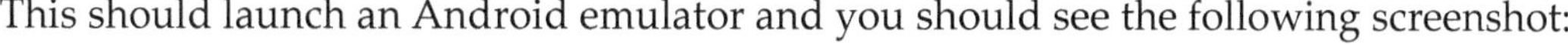

This should launch an Android emulator and you should see the following screenshot:

Congratulations! We have our first view. Well hey, no one said it would be pretty yet!

Summary

We have kicked off Part 2 with the component building, where we have laid out our root component `app.component.html` to house our primary view, where you learned about `GridLayout`, a very useful layout container.

Angular's Component decorator allowed us to easily build `TrackListComponent` as well as `PlayerControlsComponent`. We also learned how to build an Angular `Pipe` to aid our view's ability to keep our track list in order. Angular's `NgModule` taught us we need to ensure any view-related declarations needed for any components are imported properly. This Angular design pattern helps maintain module isolation as standalone units of code that can be intermixed by importing modules into each other.

We also enhanced a fair share of our services to support some of the usability we desire with our components.

Finally, we were able to take a quick peek at what we were building. Even though it's not at all pretty at this point, we can see things coming together.

In `Chapter 4`, *A prettier view with CSS*, you will learn how to use CSS to bring out the pretty from our views.

4
A prettier view with CSS

One of the many key benefits NativeScript brings to native app development is the ability to style native view components with standard CSS. You will find great support for many common and advanced properties; however, some don't have a direct correlation, whereas others are completely unique to native view layouts.

Let's take a look at how to turn our first view into something pretty amazing with a few CSS classes. You will also learn how to utilize NativeScript's core theme to provide a consistent styling framework to build on.

In this chapter, we will be covering the following topics:

- Using CSS to style views
- Understanding some of the differences between typical web styling and native styling
- Unlocking NativeScript powers with platform-specific files
- Learning how to use the nativescript-theme-core styling framework plugin
- Adjusting the status bar background color and text color on iOS and Android

It's time to get classy

Let's start by taking a look at our app's main `app.css` file inside the `App` directory:

```
/*
In NativeScript, the app.css file is where you place CSS rules that
you would like to apply to your

entire application. Check out
http://docs.nativescript.org/ui/styling for a full list of the CSS
selectors and
```

```
properties you can use to style UI components.

/*
For example, the following CSS rule changes the font size

of all UI
components that have the btn class name.
*/
.btn {
  font-size: 18;
}

/*
In many cases you may want to use the NativeScript core theme instead
of writing your own CSS rules. For a full list

of class names in the theme
refer to http://docs.nativescript.org/ui/theme.
*/
@import 'nativescript-

theme-core/css/core.light.css';
```

Out of the box, the `--ng` template hints at two options you could choose from to build out your CSS:

- Write your own custom classes
- Utilize the nativescript-theme-core styling framework plugin as your base

Let's explore the first option for a moment. Add the following after the `.btn` class:

```
.btn {
  font-size: 18;
}

.row {
  padding: 15 5;
  background-color: yellow;
}

.row .title {
  font-size: 25;
  color: #444;
  font-weight: bold;
}

Button {
```

```
  background-color: red;

color: white;
}
```

There's a number of interesting things you may pick up on right away from this simple example:

- Padding does not use the `px` suffix you may know well with web styling.
    - Don't worry, using the `px` suffix will not hurt you.
    - Starting with NativeScript 3.0, release units are supported, so you can use dp (device independent pixels) or `px` (device pixels). If no unit is specified, dp will be used. For width/height and margins, you can also use percents in CSS as a unit type.
- Various common properties (`padding`, `font size`, `font weight`, `color`, `background color`, and more) are supported. Also, shorthand margin/padding works as well, that is, padding: 15 5.
- You can use standard hex color names, such as yellow, or shorthand codes, such as #444.
- CSS scoping works as you would expect, that is, `.row .title { ....`
- Element/Tag/Component names can be styled globally.

Even though you can style by tag/component name, it is not advisable to do so. We will show you a few interesting considerations for native devices you will want to be aware of.

Now, let's open `app/modules/player/components/track-list/track-list.component.html` and add the `row` and `title` classes to our template:

```
<ListView [items]="playerService.tracks | orderBy: 'order'">
  <template let-track="item">

<GridLayout rows="auto" columns="100,*,100" class="row">
      <Button text="Record" (tap)

="record(track)" row="0" col="0"></Button>
      <Label [text]="track.name" row="0" col="1"

class="title"></Label>
      <Switch row="0" col="2"></Switch>

</GridLayout>
```

```
    </template>
  </ListView>
```

Let's quickly preview what happens with `tns run ios --emulator` and you should see the following:

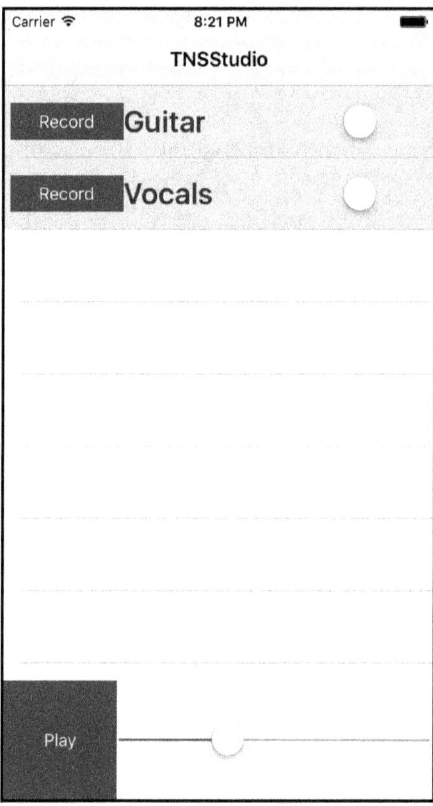

If you take a look in Android with `tns run android --emulator`, you should see the following:

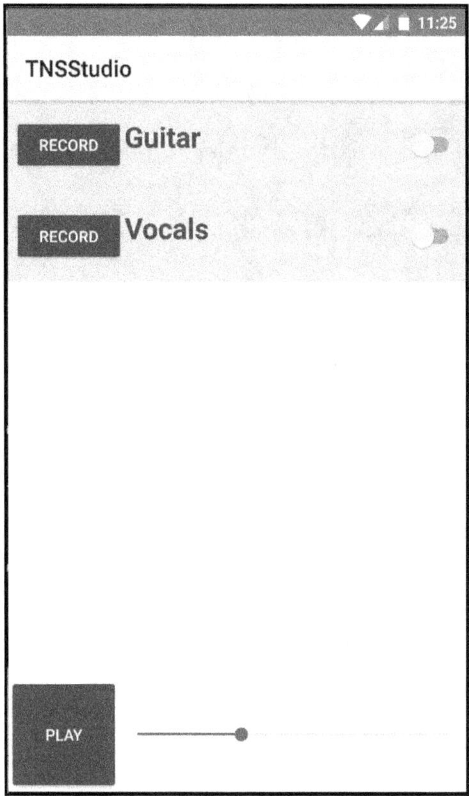

We can see, on both platforms, these styles applied consistently, while, still maintaining each platform's unique characteristics. For example, iOS maintains the flat design aesthetic across the buttons, and the switches provide that familiar iOS feel. In contrast, on Android the buttons preserve their subtle default shadows and all caps text, as well as retaining the familiar Android switches.

However, there are some subtle (potentially undesirable) differences that are important to understand and remedy. From this example, we may note the following:

1. Android's buttons have wider left/right margins than iOS.
2. Row titles are not aligned consistently. On iOS, the Label is vertically centered by default; however, on Android it's aligned to the top.

3. If you tap on the **Record** button to view the login dialog, you will also notice something quite undesirable:

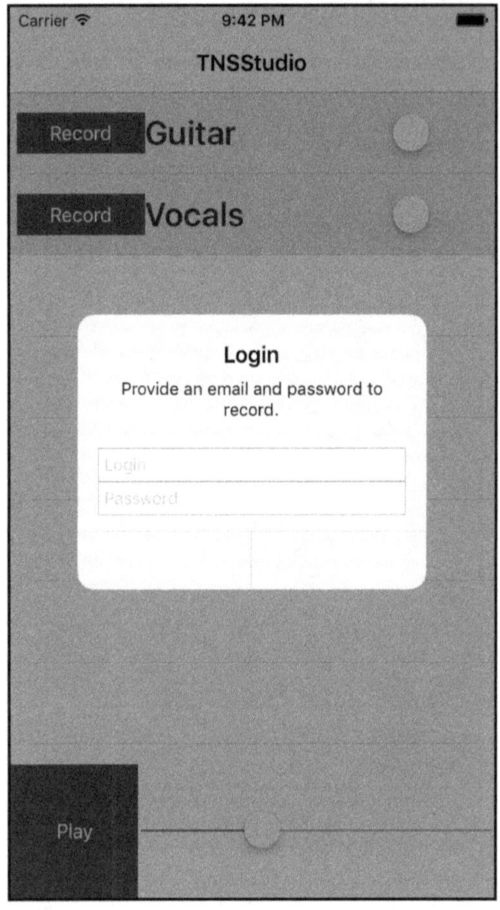

Item #3 may be the most surprising and unexpected. It exemplifies one of the main reasons it is not advisable to style Element/Tag/Component names globally. Since the native dialogs use `Buttons` by default, some of the global `Button` styles we added are bleeding into the dialog (notably `color: white`). To fix this, we can either ensure we properly scope all the component names:

```
.row Button {
  background-color: red;
  color: white;
}
```

Or better yet, just use a class name on your Buttons:

```
.row .btn {
  background-color: red;
  color: white;
}
<Button text="Record" (tap)="record(track)" row="0" col="0"

class="btn"></Button>
```

To fix item #2 (row title alignment), we can introduce a special power of NativeScript: the ability to build platform-specific files depend on which platform you are running it on. Let's create a new file, `app/common.css`, and refactor all the contents of `app/app.css` into this new file. Then, let's create two other new files, `app/app.ios.css` and `app/app.android.css` (and then delete `app.css`, since it will no longer be needed), both with the following contents:

```
@import './common.css';
```

This will ensure that our common shared styles are imported into both iOS and Android CSS. Now, we have a way to apply platform-specific styling fixes!

Let's fix that vertical alignment issue by modifying `app/app.android.css` to the following:

```
@import './common.css';

.row .title {
  vertical-align: center;
}
```

This adds the additional styling tweak for Android only to give us this now:

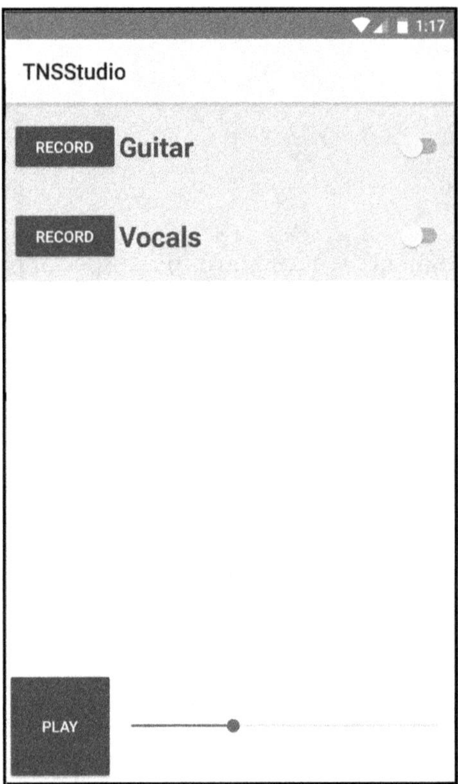

Excellent, much better.

To fix #1, we would need to apply more platform-specific tweaking if we want our buttons on both platforms to have the same margins.

At this point, you may be wondering how much tweaking you may need to do yourself to address some of these platform-specific concerns. You should be happy to know there's not an exhaustive list, but the incredibly high spirited NativeScript community worked together to create something even better, a consistent bootstrap-like core theme that provides a lot of these subtle tweaks, such as Label vertical alignment and many other subtle adjustments.

Meet the NativeScript core theme

All new NativeScript projects come with a core theme installed and are ready to use out of the box. As mentioned previously, you are provided at two options you could use to style your app. The preceding section outlined a few of the things you may run into while styling your App from scratch.

Let's take a look at Option #2: using the `nativescript-theme-core` plugin. Out of the box, this theme is built to scale and build on top of. It provides a wide assortment of utility classes for spacing, coloring, layout, colored skins, and much more. Because of the solid foundation and amazing flexibility it provides, we will build our app's styles on top of this theme.

It's worth mentioning the `nativescript-theme-` prefix was intentional as a standard, as it helps provide a common prefix to search for on `npm` to find all the NativeScript themes. It's encouraged to use the same prefix if you design and publish your own custom NativeScript theme.

Let's remove our custom styling, leaving only the core theme imported. However, instead of using the default light skin, we are going to use the dark skin. This is what our `app/common.css` file should look like now:

```
@import 'nativescript-theme-core/css/core.dark.css';
```

Now, we want to start classing our components with some of the classes that the core theme provides. You can learn a full list of classes here: `https://docs.nativescript.org/ui/theme`.

Starting with `app/app.component.html`, let's add the following classes:

```
<ActionBar title="TNSStudio" class="action-bar">
</ActionBar>
<GridLayout

rows="*, 100" columns="*" class="page">
  <track-list row="0" col="0"></track-list>
  <player-controls row="1" col="0"></player-controls>
</GridLayout>
```

The `action-bar` class ensures our skin gets applied appropriately to the header of the App, as well as providing subtle consistency adjustments for `ActionBar` on both iOS and Android.

The `page` class ensures our skin applies to the entire page. It's important this class is applied to the root layout container on any given component view.

With these two adjustments, we should now see this on iOS:

And this is on Android:

You'll notice another iOS/Android difference with `ListView`. iOS has a white background by default, whereas Android appears to have a transparent background, allowing the skin page color to show through. Let's continue classing our components with more classes from the core theme, which help address these subtleties. Open `app/modules/player/components/track-list/track-list.component.html` and add the following classes:

```
<ListView [items]="playerService.tracks | orderBy: 'order'" class="list-
group">
  <ng-

template let-track="item">
    <GridLayout rows="auto" columns="100,*,100" class="list-group-

item">
      <Button text="Record" (tap)="record(track)" row="0" col="0" class="c-
```

```
ruby"></Button>
        <Label [text]="track.name" row="0" col="1"

class="h2"></Label>
        <Switch row="0" col="2"

class="switch"></Switch>
      </GridLayout>
    </ng-template>
</ListView>
```

The parent `list-group` class helps scope everything properly down to `list-group-item`. Then, we add `c-ruby` to splash some reddish color to our **Record** buttons. There are several skinned colors that provide surnames: `c-sky`, `c-aqua`, `c-charcoal`, `c-purple`, and much more. See all of them here: `https://docs.nativescript.org/ui/theme#color-schemes`.

We then add `h2` to the Label to bring its font size up a bit. Lastly, the `switch` class helps standardize the track solo switch.

We now have this on iOS:

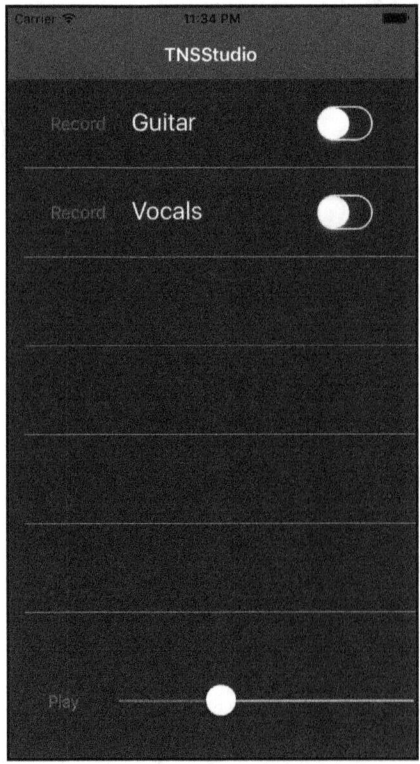

And we have this on Android:

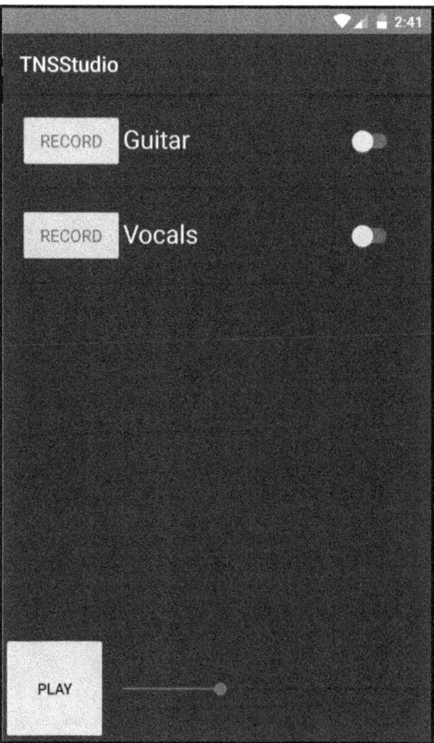

Let's move onward to our last component (for now), `player-controls`. Open `app/modules/player/components/player-controls/player-controls.component.html` and add the following:

```
<GridLayout rows="100" columns="100,*" row="1" col="0" class="p-x-10">
  <Button

[text]="playStatus" (tap)="togglePlay()" row="0" col="0" class="btn btn-
primary w-

100"></Button>
  <Slider minValue="0" [maxValue]="duration" [value]="currentTime" row="0"
col="1"

class="slider"></Slider>
</GridLayout>
```

First, we add the `p-x-10` class to add the `10` padding to only the left/right container (`GridLayout`). Then, we add `btn btn-primary w-100` to our play/pause button. The `w-100` class sets the button to have a fixed width of `100`. Then, we add the `slider` class to our Slider.

Now, things are starting to shape up on iOS:

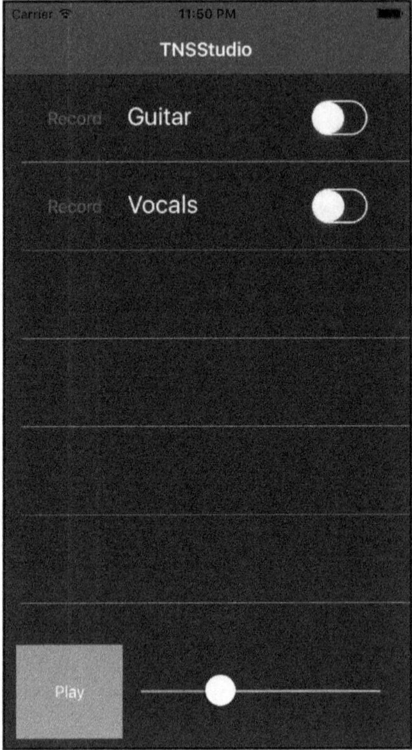

It will look as follows on Android:

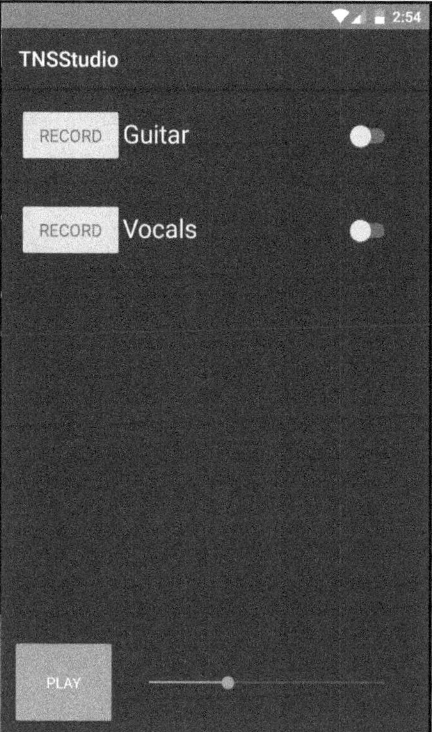

Wow, alright now, things are coming together. We will continue to polish things more as we go, but this exercise has demonstrated how quickly you can pull your styling around with the core theme by using a lot of the classes that come out of the box.

Adjusting the status bar background color and text color on iOS and Android

You may have noticed earlier that, on iOS, the status bar text is black and doesn't look very good with our dark skin. Additionally, we may want to alter Android's status bar tint color. NativeScript provides direct access to native APIs, so we can easily change these to whatever we would like. Both platforms deal with them differently, so we can conditionally alter the status bar for each.

Open `app/app.component.ts` and let's add the following:

```
// angular
import { Component } from '@angular/core';

// nativescript
import { isIOS } from 'platform';
import { topmost } from 'ui/frame';
import * as app from 'application';

// app
import { AuthService } from

'./modules/core/services';

declare var android;

@Component({
  moduleId:

module.id,
  selector: 'my-app',
  templateUrl: 'app.component.html',
})
export class AppComponent {

  constructor(
    private authService: AuthService
  ) {
    if (isIOS) {
      /**
       * 0 = black text
       * 1 = white text
       */
      topmost().ios.controller.navigationBar.barStyle = 1;


    } else {
      // adjust text to darker color
      let decorView =

app.android.startActivity.getWindow()
        .getDecorView();

decorView.setSystemUiVisibility(android.view.View.SYSTEM_UI_FLAG_LIGHT_STAT
US_BAR);
    }
  }
```

```
}
```

This will turn the iOS status bar text white:

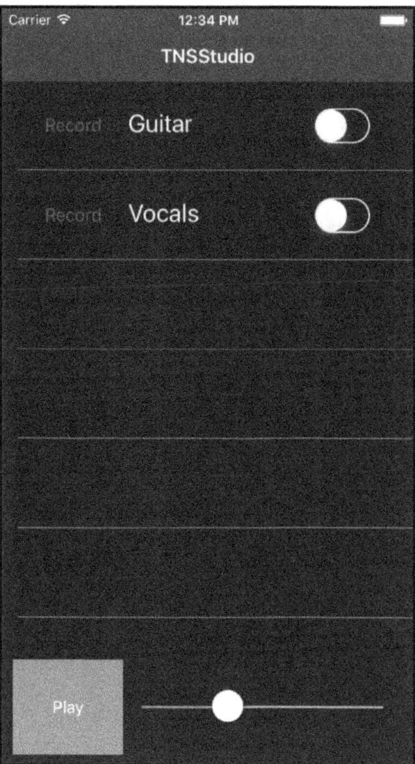

The second part of the condition adjusts Android to use dark text in the status bar:

Let's also adjust the `ActionBar` background color while we're at it for a nice touch. On iOS, the status bar background color takes on the background color of `ActionBar`, whereas on Android, the background color of the status bar must be adjusted via Android `colors.xml` in `App_Resources`. Starting with iOS, let's open `app/common.css` and add the following:

```
.action-bar {
  background-color:#101B2E;
}
```

This colors `ActionBar` as follows for iOS:

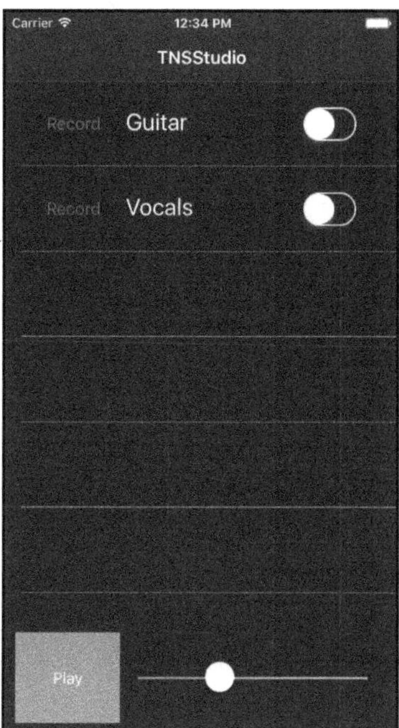

For Android, we want our status bar background to present a complimentary hue to our `ActionBar` background. To do that, we want to open `app/App_Resources/Android/values/colors.xml` and make the following adjustment:

```
<?xml version="1.0" encoding="utf-8"?>
<resources>
  <color

name="ns_primary">#F5F5F5</color>
  <color

name="ns_primaryDark">#284472</color>
  <color name="ns_accent">#33B5E5</color>

<color name="ns_blue">#272734</color>
</resources>
```

This is the final result on Android:

Summary

It's refreshing and fun to finally put a face on our app; however, we are certainly not done styling. We will continue polishing views via CSS and introduce SASS soon to refine it even more in the upcoming chapters. However, this chapter has introduced you to various considerations you will want to be aware of while styling your App via CSS.

You've learned that common CSS properties are supported, and we have also looked at differences between how iOS and Android handle certain default characteristics. The ability to have platform-specific CSS overrides is a nice benefit and special power you will want to take advantage of in your cross-platform NativeScript apps. Understanding how to control the appearance of the status bar on both platforms is essential to achieving the desired look and feel of your app.

In the next chapter, we will take a break from styling and dive into routing and navigation via lazy loading to set the stage for rounding out the general usability flow of our app. Get ready to dive into some of the more interesting Angular bits of our app.

5
Routing and Lazy Loading

Routing is essential to the solid usability flow of any app. Let's understand the key elements of routing configuration for a mobile app that takes advantage of all the flexibility Angular's router gives us.

In this chapter, we will be covering the following topics:

- Configuring the Angular Router with a NativeScript app
- Lazy loading modules by route
- Provide NSModuleFactoryLoader for Angular's NgModuleFactoryLoader
- Understanding how to use router-outlet in conjunction with page-router-outlet
- Learn how to share singleton services across multiple lazy loaded modules
- Using auth guards to protect views that require valid authentication
- Learn about `NavigationButton` to customize back mobile navigation
- Take advantage of our flexible routing setup by introducing late feature requirements

Get your kicks on Route 66

As we begin our journey down this highway full of adventure, let's start with a pit stop at our local service shop to ensure our vehicle is in tip-top shape. Take a turn into the root directory of `app` to build a new add-on to our vehicle's engine: the routing module.

Create a new routing module, `app/app.routing.ts`, with the following contents:

```
import { NgModule } from '@angular/core';
import { NativeScriptRouterModule }
  from 'nativescript-angular/router';
import { Routes } from '@angular/router';
```

```
const routes: Routes = [
  {
    path: '',
    redirectTo: '/mixer/home',
    pathMatch: 'full'
  },
  {
    path: 'mixer',
    loadChildren: () =>
require('./modules/mixer/mixer.module')['MixerModule']
  },
  {
    path: 'record',
    loadChildren: () =>
require('./modules/recorder/recorder.module')['RecorderModule']
  }
];

@NgModule({
  imports: [
    NativeScriptRouterModule.forRoot(routes)
  ],
  exports: [
    NativeScriptRouterModule
  ]
})
export class AppRoutingModule { }
```

Defining the root `''` path to redirect to a lazy loaded module provides a very flexible routing configuration, as you will see throughout this chapter. You will see a new module, `MixerModule`, which we will create momentarily. In fact, it will largely end up being what `AppComponent` is right now. Here's a list of some advantages you gain with a route configuration similar to this:

- Keeps app startup time fast by eagerly loading only the bare minimum root module configuration, then rapidly loading the first route's module lazily
- Provides us with the ability to utilize `page-router-outlet` in conjunction with `router-outlet` for a combination of master/detail navigation as well as the `clearHistory` swap page navigation
- Isolates routing configuration responsibility to the modules it concerns which scales well over time
- Allows us to target different **start pages** easily in the future if we decide to change the initial page our users are presented with

This uses `NativeScriptRoutingModule.forRoot(routes)`, since this should be considered the root of our app's routing configuration.

We also export `NativeScriptRoutingModule`, since we will be importing this `AppRoutingModule` into our root `AppModule` in a moment. This makes the routing directives available to our root module's root component.

Providing NSModuleFactoryLoader for NgModuleFactoryLoader

By default, Angular's built-in module loader uses SystemJS; however, NativeScript provides an enhanced module loader called `NSModuleFactoryLoader`. Let's provide this in our main routing module to ensure all our modules are loaded with it instead of Angular's default module loader.

Make the following modifications to `app/app.routing.ts`:

```
import { NgModule, NgModuleFactoryLoader } from '@angular/core';
import { NativeScriptRouterModule, NSModuleFactoryLoader } from
'nativescript-angular/router';

const routes: Routes = [
  {
    path: '',
    redirectTo: '/mixer/home',
    pathMatch: 'full'
  },
  {
    path: 'mixer',
    loadChildren: './modules/mixer/mixer.module#MixerModule'
  },
  {
    path: 'record',
    loadChildren: './modules/recorder/recorder.module#RecorderModule',
    canLoad: [AuthGuard]
  }
];

@NgModule({
  imports: [
    NativeScriptRouterModule.forRoot(routes)
  ],
  providers: [
    AuthGuard,
```

```
      {
        provide: NgModuleFactoryLoader,
        useClass: NSModuleFactoryLoader
      }
    ],
    exports: [
      NativeScriptRouterModule
    ]
  })
  export class AppRoutingModule { }
```

Now, we can use the standard Angular lazy loading syntax via `loadChildren` by specifying the default `NgModuleFactoryLoader` but should instead use NativeScript's enhanced `NSModuleFactoryLoader`. We won't go into what `NSModuleFactoryLoader` provides in detail, since it is explained very well here: `https://www.nativescript.org/blog/optimizing-app-loading-time-with-angular-2-lazy-loading`, and we have a lot more we want to cover in this book.

Excellent. With these upgrades in place, we can leave the service shop and continue on our journey down the highway. Let's move on to implementing our new routing setup.

Open `app/app.component.html`; cut its contents to the clipboard and replace them with the following:

```
<page-router-outlet></page-router-outlet>
```

This will be the base of our view level implementation. `page-router-outlet` allows any Component to insert itself in its place, whether it be a single flat route or one with child views of its own. It also allows other Component views to push onto the mobile nav stack, allowing master/detail mobile navigation with back history.

In order for this `page-router-outlet` directive to work, we need our root `AppModule` to import our new `AppRoutingModule`. We will also take this opportunity to remove `PlayerModule`, which was imported here before. Open `app/app.module.ts` and make the following modifications:

```
// angular
import { NgModule } from '@angular/core';

// app
import { CoreModule } from './modules/core/core.module';
import { AppRoutingModule } from './app.routing';
import { AppComponent } from './app.component';

@NgModule({
 imports: [
```

```
    CoreModule,
    AppRoutingModule
  ],
  declarations: [AppComponent],
  bootstrap: [AppComponent]
})
export class AppModule { }
```

Creating MixerModule

This module really won't be anything new, as it will serve as a relocation of what was previously our root component's view. However, it will introduce an extra nicety: the ability to define its own inner routes.

Create `app/modules/mixer/components/mixer.component.html` and paste the contents from where we had cut from the `app.component.html`:

```
<ActionBar title="TNSStudio" class="action-bar"></ActionBar><GridLayout
rows="*, 100" columns="*" class="page">
  <track-list row="0" col="0"></track-list>
  <player-controls row="1" col="0"></player-controls></GridLayout>
```

Then create a matching `app/modules/mixer/components/mixer.component.ts`:

```
import { Component } from '@angular/core';

@Component({
  moduleId: module.id,
  selector: 'mixer',
  templateUrl: 'mixer.component.html'
})
export class MixerComponent {}
```

Now, we will create `BaseComponent`, which will serve as the placeholder for not only the preceding `MixerComponent` but also any other child view components we may want to present in its place. For example, our mixer may want to allow users to pop a single track out of the mixer and into an isolated view to work with audio effects.

Create app/modules/mixer/components/base.component.ts with the following:

```
// angular
import { Component } from '@angular/core';

@Component({
 moduleId: module.id,
 selector: 'mixer-base',
 template: `<router-outlet></router-outlet>`
})
export class BaseComponent { }
```

This provides a slot to insert any child routes our mixer configures, one of which is MixerComponent itself. Since the view is just a simple router-outlet, there's really no need to create a separate templateUrl, so we have just inlined it here.

Now we are ready to implement MixerModule; create app/modules/mixer/mixer.module.ts with the following:

```
import { NgModule, NO_ERRORS_SCHEMA } from '@angular/core';
import { NativeScriptRouterModule } from
  'nativescript-angular/router';
import { Routes } from '@angular/router';

import { PlayerModule } from '../player/player.module';
import { BaseComponent } from './components/base.component';
import { MixerComponent } from
  './components/mixer.component';

const COMPONENTS: any[] = [
  BaseComponent,
  MixerComponent
]

const routes: Routes = [
  {
    path: '',
    component: BaseComponent,
    children: [
      {
        path: 'home',
        component: MixerComponent
      }
    ]
  }
];
```

```
@NgModule({
  imports: [
    PlayerModule,
    NativeScriptRouterModule.forChild(routes)
  ],
  declarations: [
    ...COMPONENTS
  ],
  schemas: [
    NO_ERRORS_SCHEMA
  ]
})
export class MixerModule { }
```

We have imported `PlayerModule` since the mixer uses components/widgets defined there (namely, `track-list` and `player-controls`). We are also utilizing the `NativeScriptRouterModule.forChild(routes)` method to indicate that these are specifically child routes. Our route configuration sets up the BaseComponent at the root `' '` path, which defines `'home'` as `MixerComponent`. If you recall, our app's `AppRoutingModule` configured the root path of our app, as follows:

```
...
{
  path: '',
  redirectTo: '/mixer/home',
  pathMatch: 'full'
},
...
```

This will route directly to `MixerComponent` here, defined as `'home'`. We could easily direct the start page to a different view by pointing `redirectTo` at a different child view of our mixer if we wanted. Since `BaseComponent` is simply a `router-outlet`, any children defined underneath the root `' '` of our mixer's routes (seen by our the overall app's routes as `'/mixer'`) will insert directly in that view slot. If you were to run this right now, you should see the same start page we had before.

Congrats! Your app's start time is now fast and you have lazily loaded your first module!

However, there's a couple of surprising things to note:

- You may notice a quick white flash before the start page appears (on iOS at least)
- You might notice the console log prints `` `Current user:` `` twice

We will address each of these issues respectively.

1. Remove the white flash after the splash screen before the start page displays.

   This is normal and is the result of the default Page background color which is white. To provide a seamless launch experience, open the `app/common.css` file and drop this global `Page` class definition to tint the background-color to the same as our `ActionBar` background-color:

   ```
   Page {
     background-color:#101B2E;
   }
   ```

   Now, there will be no more white flash and the launch of the app will appear seamless.

2. The console log prints `` `Current user:` `` twice

   Angular's dependency injector is causing this due to lazy loading.

This comes from `app/modules/core/services/auth.service.ts`, where we had a private `init` method that was being called from the service's constructor:

```
...
@Injectable()
export class AuthService {
   ...
   constructor(
     private databaseService: DatabaseService,
     private logService: LogService
   ) {
     this._init();
   }
  ...
  private _init() {
    AuthService.CURRENT_USER = this.databaseService.getItem(
      DatabaseService.KEYS.currentUser);
    this.logService.debug(`Current user: `,
      AuthService.CURRENT_USER);
    this._notifyState(!!AuthService.CURRENT_USER);
  }
  ...
}
```

Wait! What?! Does this mean `AuthService` is getting constructed twice??!!

Yes. It does. :(

I can hear the sound of the car's wheels squealing, as you veer off this highway adventure into a ditch right about now. ;)

This is most certainly a huge problem, as we absolutely intended for `AuthService` to be a globally shared Singleton that could be injected anywhere and shared to provide the current authenticated state of our app.

It is imperative we solve this right now, but let's first take a brief detour to understand why this is happening before looking at a solid solution.

Understanding Angular's Dependency Injector when lazy loading modules

Instead of restating the details, we will paraphrase directly from Angular's official documentation (`https://angular.io/docs/ts/latest/cookbook/ngmodule-faq.html#!#q-why-child-injector`), which explains this perfectly:

> *Angular adds* `@NgModule.providers` *to the application root injector unless the module is lazy loaded. For a lazy-loaded module, Angular creates a child injector and adds the module's providers to the child injector.*
>
> *This means that a module behaves differently depending on whether it's loaded during application start or lazily loaded later. Neglecting that difference can lead to adverse consequences.*
>
> *Why doesn't Angular add lazy-loaded providers to the app root injector as it does for eagerly loaded modules?*
>
> *The answer is grounded in a fundamental characteristic of the Angular dependency-injection system. An injector can add providers until it's first used. Once an injector starts creating and delivering services, its provider list is frozen; no new providers are allowed.*
>
> *When an application starts, Angular first configures the root injector with the providers of all eagerly loaded modules before creating its first component and injecting any of the provided services. Once the application begins, the app root injector is closed to new providers.*

Time passes and application logic triggers lazy loading of a module. Angular must add the lazy-loaded module's providers to an injector somewhere. It can't add them to the app root injector because that injector is closed to new providers. So Angular creates a new child injector for the lazy-loaded module context.

If we look at our root `AppModule`, we can see it imports `CoreModule`, which provides `AuthService`:

```
...
@NgModule({
  imports: [
    CoreModule,
    AppRoutingModule
  ],
  declarations: [AppComponent],
  bootstrap: [AppComponent],
  schemas: [NO_ERRORS_SCHEMA]
})
export class AppModule { }
```

If we then look at `PlayerModule`, we can see it also imports `CoreModule`, since the components of `PlayerModule` make use of the `OrderByPipe` it declares as well as several of the services it provides (that is, `AuthService`, `LogService`, and `DialogService`):

```
...
@NgModule({
  imports: [
    CoreModule
  ],
  providers: [...PROVIDERS],
  declarations: [...COMPONENTS],
  exports: [...COMPONENTS],
  schemas: [ NO_ERRORS_SCHEMA ]
})
export class PlayerModule { }
```

`PlayerModule` is now lazily loaded along with `MixerModule` due to our fancy new routing configuration. This causes Angular's dependency injector to register a new child injector for our lazily loaded `MixerModule`, which brings along `PlayerModule`, which also brings along its import of `CoreModule`, which defines those providers, including `AuthService`, `LogService`, and so on. When Angular registers `MixerModule`, it will register all the providers defined throughout the new module, including its imported modules with the new child injector, giving rise to the new instances of those services being constructed.

Angular's docs also provide a recommended setup for modules to remedy this situation, so let's paraphrase again from `https://angular.io/docs/ts/latest/cookbook/ngmodule-faq.html#!#q-module-recommendations`:

> **SharedModule**
> *Create a* `SharedModule` *with the components, directives, and pipes that you use everywhere in your app. This module should consist entirely of declarations, most of them exported. The* `SharedModule` *may re-export other widget modules, such as* `CommonModule`, `FormsModule`, *and modules with the UI controls that you use most widely.The* `SharedModule` *should not have providers for reasons explained previously. Nor should any of its imported or re-exported modules have providers. If you deviate from this guideline, know what you're doing and why. Import the* `SharedModule` *in your feature modules, both those loaded when the app starts and those you lazily load later. Create a* `CoreModule` *with providers for the singleton services you load when the application starts.*

Import `CoreModule` in the root `AppModule` only. Never import `CoreModule` in any other module.
Consider making `CoreModule` a pure service module with no declarations.

OK wow! That is an excellent recommendation. Particularly worthy of note is that very last line:

> *Consider making CoreModule a pure service module with no declarations.*

So, we already have `CoreModule`, which is great news, but we will want to make it a *pure service module with no declarations*. We will also *Import CoreModule in the root AppModule only. Never import CoreModule in any other module.* Then, we can create a new `SharedModule` to provide just *...the components, directives, and pipes that [we] use everywhere in [our] app.*

Let's create `app/modules/shared/shared.module.ts`, as follows:

```
// nativescript
import { NativeScriptModule } from 'nativescript-
angular/nativescript.module';

// angular
import { NgModule, NO_ERRORS_SCHEMA } from '@angular/core';

// app
import { PIPES } from './pipes';

@NgModule({
```

```
  imports: [
    NativeScriptModule
  ],
  declarations: [
    ...PIPES
  ],
  exports: [
    NativeScriptModule,
    ...PIPES
  ],
  schemas: [ NO_ERRORS_SCHEMA ]
})
export class SharedModule {}
```

For `PIPES`, we are just moving the pipes directory from `app/modules/core` to the `app/modules/shared` folder. Now, `SharedModule` is the one we can be free to import across several different modules that need any pipes or future shared components/directives it may provide. It will not define any service providers as mentioned by this suggestion:

> `SharedModule` *should not have providers for reasons explained previously, nor should any of its imported or re-exported modules have providers.*

We can then adjust `CoreModule` (located in `app/modules/core/core.module.ts`) as follows to be a pure service module with no declarations:

```
// nativescript
import { NativeScriptModule } from 'nativescript-
angular/nativescript.module';
import { NativeScriptFormsModule } from 'nativescript-angular/forms';
import {NativeScriptHttpModule } from 'nativescript-angular/http';
// angular
import { NgModule, Optional, SkipSelf } from '@angular/core';

// app
import { PROVIDERS } from './services';

const MODULES: any[] = [
  NativeScriptModule,
  NativeScriptFormsModule,
  NativeScriptHttpModule
];

@NgModule({
  imports: [
    ...MODULES
  ],
```

```
  providers: [
    ...PROVIDERS
  ],
  exports: [
    ...MODULES
  ]
})
export class CoreModule {
  constructor (
    @Optional() @SkipSelf() parentModule: CoreModule) {
    if (parentModule) {
      throw new Error(
        'CoreModule is already loaded. Import it in the AppModule only');
    }
  }
}
```

This module now only defines providers as the collection containing `AuthService`, `DatabaseService`, `DialogService`, and `LogService`, all of which we created earlier in the book, and we want to ensure they are true Singletons used across our app, whether they are used in lazy loaded modules or not.

Why do we use the `...PROVIDERS` spread notation instead of just assigning the collection directly?
For scalability reasons. In the future, if we need to add an additional provider or override a provider, we can do so simply by just adding to the collection right in the module. The same goes for imports and exports.

We also take this opportunity to import some additional modules that we want to ensure are also used globally throughout the app. `NativeScriptModule`, `NativeScriptFormsModule`, and `NativeScriptHttpModule` are all essential modules that override certain web APIs from Angular's various providers out-of-the-box to enhance our app with native APIs. For example, instead of the app using `XMLHttpRequest` (which is a web API), it will use native HTTP APIs made available on both iOS and Android for the ultimate networking performance. We ensure we export these as well so our root module no longer needs to import them and can instead just import this `CoreModule`.

Lastly, we define a constructor that will help safeguard us in the future from accidentally importing this `CoreModule` into other lazily loaded modules.

We don't know yet if `PlayerService` provided by `PlayerModule` will be needed by `RecorderModule`, which also will be lazily loaded. If that comes up in the future, we can also refactor `PlayerService` into `CoreModule` to ensure it's a true Singleton shared across our entire app. For now, we will just leave it where it is as part of `PlayerModule`.

Let's now make our final adjustments to our other modules based on what we have done to tighten everything down.

The `app/modules/player/player.module.ts` file should now look like this:

```
// angular
import { NgModule, NO_ERRORS_SCHEMA } from '@angular/core';

// app
import { SharedModule } from '../shared/shared.module';
import { COMPONENTS } from './components';
import { PROVIDERS } from './services';

@NgModule({
  imports: [ SharedModule ],
  providers: [ ...PROVIDERS ],
  declarations: [ ...COMPONENTS ],
  exports: [
    SharedModule,
    ...COMPONENTS
  ],
  schemas: [ NO_ERRORS_SCHEMA ]
})
export class PlayerModule { }
```

The `app/modules/recorder/recorder.module.ts` file should now look like this:

```
// angular
import { NgModule, NO_ERRORS_SCHEMA } from '@angular/core';

// app
import { SharedModule } from '../shared/shared.module';
import { PROVIDERS } from './services';

@NgModule({
 imports: [ SharedModule ],
 providers: [ ...PROVIDERS ],
 schemas: [ NO_ERRORS_SCHEMA ]
})
export class RecorderModule { }
```

Notice we now import `SharedModule` instead of `CoreModule`. This provides us with the ability to share directives, components, and pipes (essentially anything that would be in the declarations portion of the module) across the entire app by importing that `SharedModule`.

Our root `AppModule` at `app/app.module.ts` stays the same:

```
// angular
import { NgModule } from '@angular/core';

// app
import { CoreModule } from './modules/core/core.module';
import { AppRoutingModule } from './app.routing';
import { AppComponent } from './app.component';

@NgModule({
  imports: [
    CoreModule,
    AppRoutingModule
  ],
  declarations: [ AppComponent ],
  bootstrap: [ AppComponent ]
})
export class AppModule { }
```

Any module (lazy loaded or not) can still inject any services that `CoreModule` provides, since the root `AppModule` now imports that `CoreModule`. This allows Angular's root injector to construct the services provided by `CoreModule` exactly once. Then, any time those services are injected anywhere (*in a lazily loaded module or not*), Angular will first ask the parent injector (in the case of a lazy loaded module, it would be the child injector) for that service and, if not found there, it will ask the next parent making its way to the root injector, eventually, where those Singletons are provided.

Well, we've had an amazing time in this desert of a town. Let's cruise on down the highway to the ultra secure Area 51, where modules can be locked away for years unless proper authorization is presented.

Creating AuthGuard for RecorderModule

One of our app's requirements is that recording features should be locked away and inaccessible until a user is authenticated. This provides us with the ability to have a user base and potentially introduce paid features down the road if we so desire.

Angular provides the ability to insert guards on our routes, which would only activate under certain conditions. This is exactly what we need to implement this feature requirement, since we have isolated the `'/record'` route to lazily load `RecorderModule`, which will contain all the recording features. We want to only allow access to that `'/record'` route if the user is authenticated.

Let's create `app/guards/auth-guard.service.ts` in a new folder for scalability, since we could grow and create other guards as necessary here:

```
import { Injectable } from '@angular/core';
import { Route, CanActivate, CanLoad } from '@angular/router';
import { AuthService } from '../modules/core/services/auth.service';

@Injectable()
export class AuthGuard implements CanActivate, CanLoad {

  constructor(private authService: AuthService) { }

  canActivate(): Promise<boolean> {
    return new Promise((resolve, reject) => {
      if (this._isAuth()) {
        resolve(true);
      } else {
        // login sequence to continue prompting
        let promptSequence = (usernameAttempt?: string) => {
          this.authService.promptLogin(
            'Authenticate to record.',
            usernameAttempt
          ).then(() => {
            resolve(true);
          }, (usernameAttempt) => {
            if (usernameAttempt === false) {
              // user canceled prompt
              resolve(false);
            } else {
              // initiate sequence again
              promptSequence(usernameAttempt);
            }
          });
        };
        // start login prompt sequence
        // require auth before activating
        promptSequence();
      }
    });
  }
```

```
  canLoad(route: Route): Promise<boolean> {
    // reuse same logic to activate
    return this.canActivate();
  }

  private _isAuth(): boolean {
    // just get the latest value from our BehaviorSubject
    return this.authService.authenticated$.getValue();
  }
}
```

We are able to take advantage of `BehaviorSubject` of `AuthService` to grab the latest value using `this.authService.authenticated$.getValue()` to determine the auth state. We use this to immediately activate the route via the `canActivate` hook (or load the module via the `canLoad` hook) if the user is authenticated. Otherwise, we display the login prompt via the service's method, but this time we wrap it in a reprompt sequence, which will continue to prompt on failed attempts until a successful authentication, or ignore it if the user cancels the prompt.

For the book, we aren't wiring up to any backend service to do any real authentication with a service provider. We will leave that part up to you in your own app. We will just be persisting the e-mail and password you enter into the login prompt as a valid user after doing very simple validation on the input.

Notice that `AuthGuard` is an Injectable service like other services, so we will want to make sure it is added to the providers metadata of `AppRoutingModule`. We can now guard our route with the following highlighted modifications to `app/app.routing.ts` to use it:

```
...
import { AuthGuard } from './guards/auth-guard.service';

const routes: Routes = [
  ...
  {
    path: 'record',
    loadChildren:
      './modules/recorder/recorder.module#RecorderModule',
    canLoad: [AuthGuard]
  }
];

@NgModule({
  ...
  providers: [
    AuthGuard,
```

```
    ...
  ],
  ...
})
export class AppRoutingModule { }
```

To try this out, we need to add child routes to our `RecorderModule`, since we have not done that yet. Open `app/modules/recorder/recorder.module.ts` and add the following highlighted sections:

```
// nativescript
import { NativeScriptModule } from 'nativescript-
angular/nativescript.module';
import { NativeScriptRouterModule } from 'nativescript-angular/router';

// angular
import { NgModule, NO_ERRORS_SCHEMA } from '@angular/core';
import { Routes } from '@angular/router';

// app
import { SharedModule } from '../shared/shared.module';
import { PROVIDERS } from './services';
import { RecordComponent } from './components/record.component';

const COMPONENTS: any[] = [
  RecordComponent
]

const routes: Routes = [
  {
    path: '',
    component: RecordComponent
  }
];

@NgModule({
  imports: [
    SharedModule,
    NativeScriptRouterModule.forChild(routes)
  ],
  declarations: [ ...COMPONENTS ],
  providers: [ ...PROVIDERS ],
  schemas: [ NO_ERRORS_SCHEMA ]
})
export class RecorderModule { }
```

We now have a proper child route configuration that will display the single RecordComponent when the user navigates to the '/record' path. We won't show the details of RecordComponent, as you can refer to the Chapter 5, *Routing and Lazy Loading* branch on the repo for the book. However, it is just a stubbed out component at this point inside app/modules/recorder/components/record.component.html, which just shows a simple label, so we can try this out.

Lastly, we need a button that will route to our '/record' path. If we look back at our original sketch, we wanted a Record button to display in the top right corner of ActionBar, so let's implement that now.

Open app/modules/mixer/components/mixer.component.html and add the following:

```
<ActionBar title="TNSStudio" class="action-bar">
  <ActionItem nsRouterLink="/record" ios.position="right">
    <Button text="Record" class="action-item"></Button>
  </ActionItem>
</ActionBar>
<GridLayout rows="*, 100" columns="*" class="page">
  <track-list row="0" col="0"></track-list>
  <player-controls row="1" col="0"></player-controls>
</GridLayout>
```

Now, if we were to run this in the iOS Simulator, we would notice that our Record button in ActionBar does not do anything! This is because MixerModule only imports the following:

```
@NgModule({
  imports: [
    PlayerModule,
    NativeScriptRouterModule.forChild(routes)
  ],
  ...
})
export class MixerModule { }
```

The NativeScriptRouterModule.forChild(routes) method just configures the routes but does not make various routing directives, such as nsRouterLink, available to our components.

Since you learned earlier that SharedModule should be used to declare various directives, components, and pipes you want to share throughout your modules (lazy loaded or not), this is a perfect opportunity to take advantage of that.

Open `app/modules/shared/shared.module.ts` and make the following highlighted modifications:

```
...
import { NativeScriptRouterModule } from 'nativescript-angular/router';
...

@NgModule({
  imports: [
    NativeScriptModule,
    NativeScriptRouterModule
  ],
  declarations: [
    ...PIPES
  ],
  exports: [
    NativeScriptModule,
    NativeScriptRouterModule,
    ...PIPES
  ],
  schemas: [NO_ERRORS_SCHEMA]
})
export class SharedModule { }
```

Now, back in `MixerModule`, we can adjust the imports to use `SharedModule`:

```
...
import { SharedModule } from '../shared/shared.module';

@NgModule({
  imports: [
    PlayerModule,
    SharedModule,
    NativeScriptRouterModule.forChild(routes)
  ],
  ...
})
export class MixerModule { }
```

This ensures all the directives exposed via `NativeScriptRouterModule` are now included and available for use in `MixerModule` by utilizing our app-wide `SharedModule`.

Running our app again, we now see the login prompt when we tap the **Record** button in `ActionBar`. If we enter a properly formatted e-mail address and any password, it will persist the details, log us in, and display `RecordComponent` as follows on iOS:

You might notice something rather interesting. `ActionBar` changed from the background color we assigned via CSS and the button color now displays the default blue color. This is because `RecordComponent` does not define `ActionBar`; therefore, it is reverting to a default styled `ActionBar` with a default back button, which takes on the title of the page it just navigated from. The `'/record'` route is also using the ability of `page-router-outlet` to push components onto the mobile navigation stack. `RecordComponent` is animated into view while allowing the user to choose the top left button to navigate back (to pop the navigation history back one).

To fix `ActionBar`, let's just add `ActionBar` to the `RecordComponent` view with a custom `NavigationButton` (a `NativeScript` view component simulating a mobile device's default back navigation button). We can make the adjustments to `app/modules/record/components/record.component.html`:

```
<ActionBar title="Record" class="action-bar">
  <NavigationButton text="Back"
    android.systemIcon="ic_menu_back">
  </NavigationButton>
</ActionBar>
<StackLayout class="p-20">
  <Label text="TODO: Record" class="h1 text-center"></Label>
</StackLayout>
```

Now, this looks a lot better:

If we run this on Android and log in using any e-mail/password combo to persist a user, it will display the same `RecordComponent` view; however, you will notice another interesting detail. We have set up Android to display a standard back arrow system icon as `NavigationButton`, but when tapping that arrow, it does not do anything. Android's default behavior relies on the device's physical hardware back button next to the home button. However, we can provide a consistent experience by just adding a tap event to `NavigationButton`, so both iOS and Android react the same to tapping the back button. Make the following modification to the template:

```
<ActionBar title="Record" icon="" class="action-bar">
  <NavigationButton (tap)="back()" text="Back"
    android.systemIcon="ic_menu_back">
  </NavigationButton>
</ActionBar>
<StackLayout class="p-20">
  <Label text="TODO: Record" class="h1 text-center"></Label>
</StackLayout>
```

Then, we can implement the `back()` method in `app/modules/recorder/components/record.component.ts` using `NativeScript` for Angular's `RouterExtensions` service:

```
// angular
import { Component } from '@angular/core';
import { RouterExtensions } from 'nativescript-angular/router';

@Component({
 moduleId: module.id,
 selector: 'record',
 templateUrl: 'record.component.html'
})
export class RecordComponent {

  constructor(private router: RouterExtensions) { }

  public back() {
    this.router.back();
  }
}
```

Now, Android's back button can be tapped to navigate back in addition to the hardware back button. iOS simply ignores the tap event handler, since it uses the default native behavior for `NavigationButton`. Pretty nice. Here is how `RecordComponent` looks on Android:

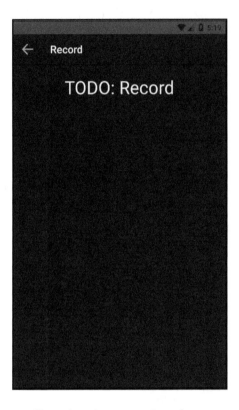

We will implement a nice recording view in upcoming chapters.

We are surely cruising down Route 66 by now!

We have implemented lazily loaded routes, provided `AuthGuard` to protect unauthorized use of our app's recording features, and learned a ton in the process. *However, we've just realized we are missing a very important feature late in the game.* We need a way to work on several different mixes over time. By default, our app may launch the last opened mix, but we would like to create new mixes (let's consider them **compositions**) and record entirely new mixes of individual tracks as separate compositions. We need a new route to display these compositions that we can name appropriately, so we can jump back and forth and work on different material.

Handling late feature requirements – managing compositions

It's time to deal with unexpected traffic along Route 66. We have encountered a late feature requirement, realizing we need a way to manage any number of different mixes so we can work on different material over time. We could refer to each mix as a composition of audio tracks.

The good news is we have spent a reasonable amount of time engineering a scalable architecture and we are about to reap the fruits of our labor. Responding to late feature requirements now becomes a rather enjoyable Sunday stroll around the neighborhood. Let's show off the strengths of our app's architecture by taking a moment to work on this new feature.

Let's start by defining a new route for a new `MixListComponent` we will create. Open `app/modules/mixer/mixer.module.ts` and make the following highlighted modifications:

```
...
import { MixListComponent } from './components/mix-list.component';
import { PROVIDERS } from './services';

const COMPONENTS: any[] = [
  BaseComponent,
  MixerComponent,
  MixListComponent
]

const routes: Routes = [
  {
    path: '',
    component: BaseComponent,
    children: [
      {
        path: 'home',
        component: MixListComponent
      },
      {
        path: ':id',
        component: MixerComponent
      }
    ]
  }
];
```

```
@NgModule({
   ...
   providers: [
     ...PROVIDERS
   ]
})
export class MixerModule { }
```

We are switching up our initial strategy of presenting `MixerComponent` as the home start page, but instead we are going to create a new `MixListComponent` in a moment to represent the `'home'` start page, which will be a listing of all the compositions we are working on. We could still have the `MixListComponent` auto select the last selected composition on the app launch for convenience later. We have now defined `MixerComponent` as a parameterized route, since it will always represent one of our working compositions identified by the `':id'` param routes, which will resolve to a route looking like `'/mixer/1'` for example. We have also imported `PROVIDERS`, which we will create in a moment.

Let's modify `DatabaseService` provided by `CoreModule` to help provide a constant persistence key for our new data needs. We will want to persist user created compositions stored via this constant key name. Open `app/modules/core/services/database.service.ts` and make the following highlighted modifications:

```
...
interface IKeys {
  currentUser: string;
  compositions: string;
}

@Injectable()
export class DatabaseService {

  public static KEYS: IKeys = {
    currentUser: 'current-user',
    compositions: 'compositions'
  };
...
```

Let's also create a new data model to represent our compositions. Create `app/modules/shared/models/composition.model.ts`:

```
import { ITrack } from './track.model';

export interface IComposition {
  id: number;
```

```
  name: string;
  created: number;
  tracks: Array<ITrack>;
  order: number;
}
export class CompositionModel implements IComposition {
  public id: number;
  public name: string;
  public created: number;
  public tracks: Array<ITrack> = [];
  public order: number;

  constructor(model?: any) {
    if (model) {
      for (let key in model) {
        this[key] = model[key];
      }
    }
    if (!this.created) this.created = Date.now();
    // if not assigned, just assign a random id
    if (!this.id)
      this.id = Math.floor(Math.random() * 100000);
  }
}
```

Then, holding strong to our conventions, open `app/modules/shared/models/index.ts` and re-export this new model:

```
export * from './composition.model';
export * from './track.model';
```

We can now use this new model and database key in a new data service on which to build this new feature. Create `app/modules/mixer/services/mixer.service.ts`:

```
// angular
import { Injectable } from '@angular/core';

// app
import { ITrack, IComposition, CompositionModel } from
'../../shared/models';
import { DatabaseService } from '../../core/services/database.service';
import { DialogService } from '../../core/services/dialog.service';

@Injectable()
export class MixerService {

  public list: Array<IComposition>;
```

```
constructor(
  private databaseService: DatabaseService,
  private dialogService: DialogService
) {
  // restore with saved compositions or demo list
  this.list = this._savedCompositions() ||
    this._demoComposition();
}

public add() {
  this.dialogService.prompt('Composition name:')
    .then((value) => {
      if (value.result) {
        let composition = new CompositionModel({
          id: this.list.length + 1,
          name: value.text,
          order: this.list.length // next one in line
        });
        this.list.push(composition);
        // persist changes
        this._saveList();
      }
    });
}

public edit(composition: IComposition) {
  this.dialogService.prompt('Edit name:', composition.name)
    .then((value) => {
      if (value.result) {
        for (let comp of this.list) {
          if (comp.id === composition.id) {
            comp.name = value.text;
            break;
          }
        }
        // re-assignment triggers view binding change
        // only needed with default change detection
        // when object prop changes in collection
        // NOTE: we will use Observables in ngrx chapter
        this.list = [...this.list];
        // persist changes
        this._saveList();
      }
    });
}

private _savedCompositions(): any {
  return this.databaseService
```

```
      .getItem(DatabaseService.KEYS.compositions);
  }

  private _saveList() {
    this.databaseService
      .setItem(DatabaseService.KEYS.compositions, this.list);
  }

  private _demoComposition(): Array<IComposition> {
    // Starter composition to demo on first launch
    return [
      {
        id: 1,
        name: 'Demo',
        created: Date.now(),
        order: 0,
        tracks: [
          {
            id: 1,
            name: 'Guitar',
            order: 0
          },
          {
            id: 2,
            name: 'Vocals',
            order: 1
          }
        ]
      }
    ]
  }
}
```

We now have a service that will provide a list to bind our view to display the user's saved compositions. It also provides a way to add and edit compositions and seed the first app launch with a demo composition for a good first-time user experience (*we will add actual tracks to the demo later*).

In keeping with our conventions, let's also add `app/modules/mixer/services/index.ts`, as follows, which we illustrated being imported in `MixerModule` a moment ago:

```
import { MixerService } from './mixer.service';

export const PROVIDERS: any[] = [
  MixerService
];
```

```
export * from './mixer.service';
```

Let's now create `app/modules/mixer/components/mix-list.component.ts` to consume and project our new data service:

```
// angular
import { Component } from '@angular/core';

// app
import { MixerService } from '../services/mixer.service';

@Component({
  moduleId: module.id,
  selector: 'mix-list',
  templateUrl: 'mix-list.component.html'
})
export class MixListComponent {

  constructor(public mixerService: MixerService) { }
}
```

And, for the view template, `app/modules/mixer/components/mix-list.component.html`:

```
<ActionBar title="Compositions" class="action-bar">
  <ActionItem (tap)="mixerService.add()"
    ios.position="right">
    <Button text="New" class="action-item"></Button>
  </ActionItem>
</ActionBar>
<ListView [items]="mixerService.list | orderBy: 'order'"
  class="list-group">
  <ng-template let-composition="item">
    <GridLayout rows="auto" columns="100,*,auto"
      class="list-group-item">
      <Button text="Edit" row="0" col="0"
        (tap)="mixerService.edit(composition)"></Button>
      <Label [text]="composition.name"
        [nsRouterLink]="['/mixer', composition.id]"
        class="h2" row="0" col="1"></Label>
      <Label [text]="composition.tracks.length"
        class="text-right" row="0" col="2"></Label>
    </GridLayout>
  </ng-template>
</ListView>
```

This will render our list of `MixerService` user-saved compositions to the view and, when we first launch the app, it will have been seeded with one sample **Demo** composition preloaded with two recordings, so the user can play around. Here is how things look on iOS upon first launch now:

We can create new compositions and edit the names of existing ones. We can also tap the composition's name to view `MixerComponent`; however, we need to adjust the component to grab the route `':id'` param and wire its view into the selected composition. Open `app/modules/mixer/components/mixer.component.ts` and add the highlighted sections:

```
// angular
import { Component, OnInit, OnDestroy } from '@angular/core';
import { ActivatedRoute } from '@angular/router';
import { Subscription } from 'rxjs/Subscription';

// app
import { MixerService } from '../services/mixer.service';
import { CompositionModel } from '../../shared/models';

@Component({
 moduleId: module.id,
 selector: 'mixer',
 templateUrl: 'mixer.component.html'
})
export class MixerComponent implements OnInit, OnDestroy {

  public composition: CompositionModel;
  private _sub: Subscription;

  constructor(
    private route: ActivatedRoute,
    private mixerService: MixerService
  ) { }

  ngOnInit() {
    this._sub = this.route.params.subscribe(params => {
      for (let comp of this.mixerService.list) {
        if (comp.id === +params['id']) {
          this.composition = comp;
          break;
        }
      }
    });
  }

  ngOnDestroy() {
    this._sub.unsubscribe();
  }
}
```

We can inject Angular's `ActivatedRoute` to subscribe to the route's params, which give us access to `id`. Because it will come in as a String by default, we use `+params['id']` to convert it to a number when we locate the composition in our service's list. We assign a local reference to the selected `composition`, which now allows us to bind to it in the view. While we're at it, we will also add a Button labeled `List` for now in `ActionBar` to navigate back to our compositions (*later, we will implement font icons to display in their place*). Open `app/modules/mixer/components/mixer.component.html` and make the following highlighted modifications:

```
<ActionBar [title]="composition.name" class="action-bar">
  <ActionItem nsRouterLink="/mixer/home">
    <Button text="List" class="action-item"></Button>
  </ActionItem>
  <ActionItem nsRouterLink="/record" ios.position="right">
    <Button text="Record" class="action-item"></Button>
  </ActionItem>
</ActionBar>
<GridLayout rows="*, 100" columns="*" class="page">
  <track-list [tracks]="composition.tracks" row="0" col="0"></track-list>
  <player-controls row="1" col="0"></player-controls>
</GridLayout>
```

This allows us to display the selected composition's name in the title of `ActionBar` as well as pass its tracks to `track-list`. We need to add `Input` to `track-list`, so it renders the composition's tracks instead of the dummy data it's bound to now. Let's open `app/modules/player/components/track-list/track-list.component.ts` and add an `Input`:

```
...
export class TrackListComponent {

 @Input() tracks: Array<ITrack>;

 ...
}
```

Previously, the `TrackListComponent` view was bound to `playerService.tracks`, so let's adjust the view template for the component at `app/modules/player/components/track-list/track-list.component.html` to bind to our new `Input`, which will now represent the tracks in the user's actual selected composition:

```
<ListView [items]="tracks | orderBy: 'order'" class="list-group">
  <template let-track="item">
    <GridLayout rows="auto" columns="100,*,100" class="list-group-item">
```

```
        <Button text="Record" (tap)="record(track)" row="0" col="0" class="c-
ruby"></Button>
        <Label [text]="track.name" row="0" col="1" class="h2"></Label>
        <Switch [checked]="track.solo" row="0" col="2"
class="switch"></Switch>
      </GridLayout>
    </template>
</ListView>
```

We now have the following sequence in our app to meet the needs of this late feature requirement and we did it in just a few pages of material here:

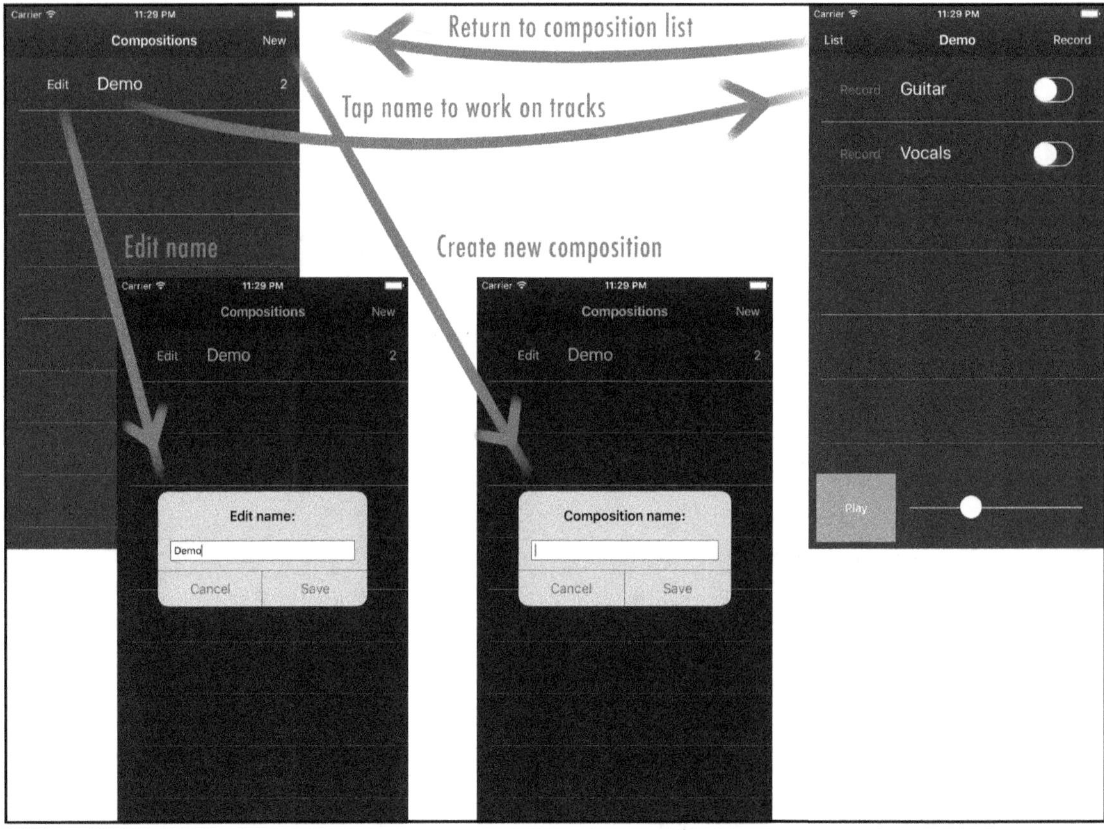

And it works exactly the same on Android while retaining its unique native characteristics.

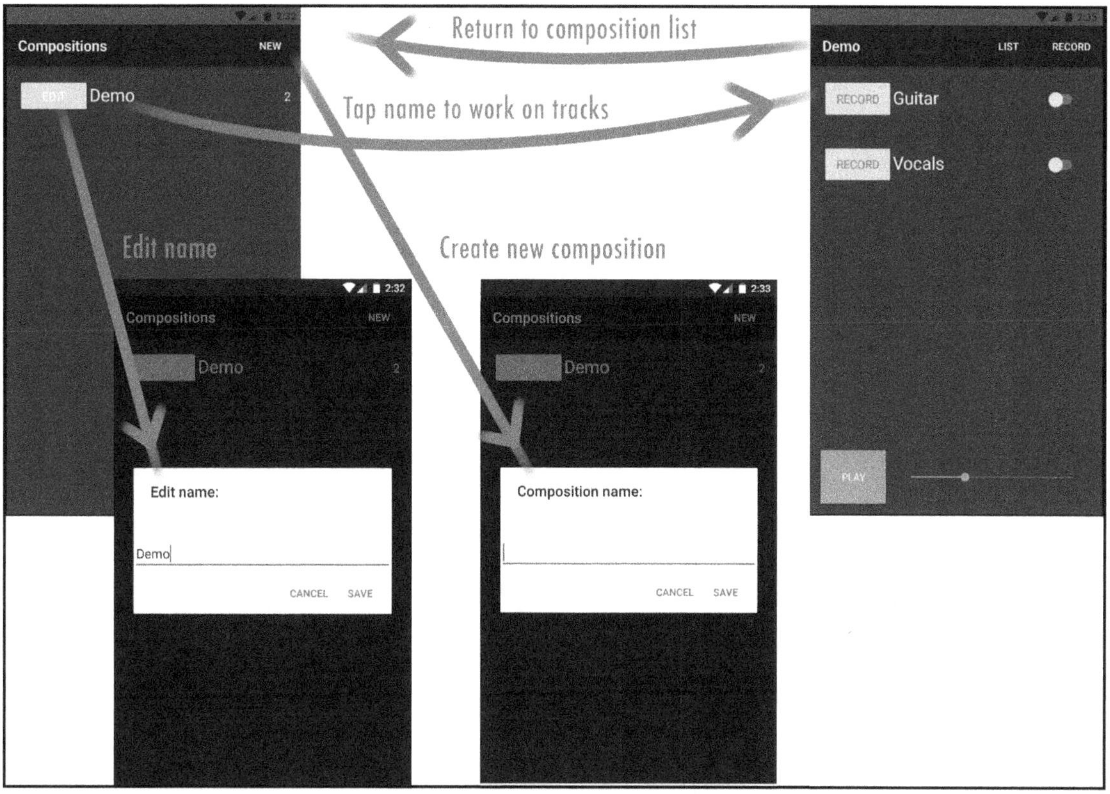

You might notice, however, that `ActionBar` on Android defaults to all `ActionItem` on the right-hand side. One last trick we want to show you quickly is the ability for platform-specific view templates. Oh and don't worry about those ugly Android buttons; we will integrate font icons later for those.

Create platform-specific view templates wherever you see fit. Doing so will help you dial views for each platform where necessary and make them highly maintainable.

Let's create `app/modules/mixer/components/action-bar/action-bar.component.ts`:

```
// angular
import { Component, Input } from '@angular/core';

@Component({
  moduleId: module.id,
  selector: 'action-bar',
  templateUrl: 'action-bar.component.html'
})
export class ActionBarComponent {

  @Input() title: string;
}
```

You can then create an iOS-specific view template: `app/modules/mixer/components/action-bar/action-bar.component.ios.html`:

```
<ActionBar [title]="title" class="action-bar">
  <ActionItem nsRouterLink="/mixer/home">
    <Button text="List" class="action-item"></Button>
  </ActionItem>
  <ActionItem nsRouterLink="/record" ios.position="right">
    <Button text="Record" class="action-item"></Button>
  </ActionItem>
</ActionBar>
```

And an Android-specific view template: `app/modules/mixer/components/action-bar/action-bar.component.android.html`:

```
<ActionBar class="action-bar">
  <GridLayout rows="auto" columns="auto,*,auto" class="action-bar">
    <Button text="List" nsRouterLink="/mixer/home" class="action-item"
row="0" col="0"></Button>
    <Label [text]="title" class="action-bar-title text-center" row="0"
col="1"></Label>
    <Button text="Record" nsRouterLink="/record" class="action-item"
row="0" col="2"></Button>
  </GridLayout>
</ActionBar>
```

Then we can use it in `app/modules/mixer/components/mixer.component.html`:

```
<action-bar [title]="composition.name"></action-bar>
<GridLayout rows="*, 100" columns="*" class="page">
  <track-list [tracks]="composition.tracks" row="0" col="0"></track-list>
  <player-controls row="1" col="0"></player-controls>
</GridLayout>
```

Just ensure you add it to the `COMPONENTS` of `MixerModule` in `app/modules/mixer/mixer.module.ts`:

```
...
import { ActionBarComponent } from './components/action-bar/action-
bar.component';
...

const COMPONENTS: any[] = [
  ActionBarComponent,
  BaseComponent,
  MixerComponent,
  MixListComponent
];
...
```

Voila!

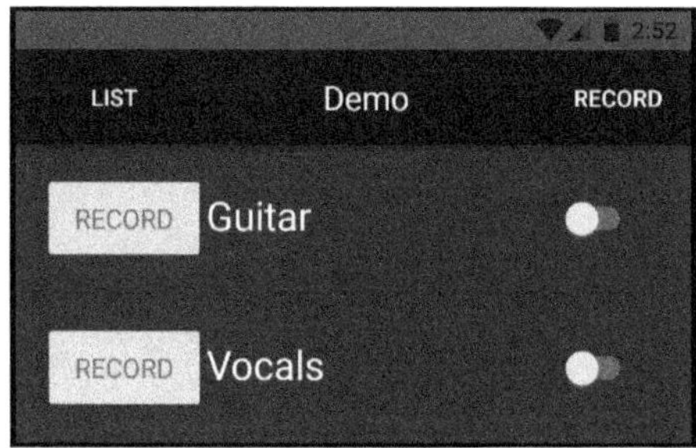

Summary

We have arrived at the end of this amazing journey down Route 66 and hope you feel as exhilarated as we do. This chapter has presented some interesting Angular concepts, including route configuration with lazy loaded modules to keep the app startup time fast; building a custom module loader using native file handling APIs; combining the flexibility of `router-outlet` with NativeScript's `page-router-outlet`; gaining control and understanding of Singleton services with lazy loaded modules; guarding routes dependent on authorized access; and working on late feature requirements to show off our wonderfully scalable app design.

This chapter rounds out the general usability flow of our app and, at this point, we are ready to venture into the core competency of our app: **Audio Handling via iOS and Android's rich native APIs**.

Before delving into the thick of things, in the next chapter we will take a brief moment to inspect NativeScript's various `tns` command-line arguments to run our app to lock in a thorough education of the tool belt we can now bring to work.

6
Running the App on iOS and Android

There are a couple of ways to build, run, and start working with NativeScript applications. We will cover command-line tools, as they are currently the most supported method, and the best way to do anything with any NativeScript project.

To simplify things for our understanding, we will work through the commands that we will use frequently first, then we will cover the rest of the commands that aren't as frequently used. So, let's begin and work through the commands that you will want to know.

In this chapter, we will cover the following topics:

- How to run an application
- How to start the Debugger
- How to build an application for deployment
- How to start the testing framework
- How to run a NativeScript diagnostic
- All about Android Keystores

Taking command...

The first command we will cover is the one you will use every time you

start your app. To make things simpler, I will use `<platform>` to mean iOS, Android, or--when it is finally supported--Windows.

tns run <platform>

The `tns run <platform>` command will automatically build your app and sync it up to the devices and emulators. It will do all the heavy-lifting to try and make your app be in a running state on the device, and then it will launch the app. This command has changed over the years and has now become a fairly smart command that will automatically make certain choices to simplify your development life. One of the cool features of this command is that it will sync your application to all running and connected devices. If you have five different devices hooked up, all five of them will receive the changes. This only works per each platform, but you can run `tns run ios` in one and `tns run android` in another command window, and then any changes will automatically be synced to all devices connected to your machine. As you may imagine, this is very useful during the testing and cleaning up phase to make sure that everything continues to look good on different phones and tablets. If you have no physical devices hooked up to your computer, it will automatically launch an emulator for you.

Normally, since the app already exists on the devices, it will just do a quick live sync of the changed files. This is a very fast process, as it just transfers all the changes in your files from your own `app` folder to all the connected devices, and then starts the app. This process is, in the majority of situations, a really good thing. However, `tns run <platform>` will not always automatically detect any changes to your `node_modules` folder, for example, when you upgrade a plugin. If this is the case, you will need to cancel the current running `tns run` and then start a new `tns run`. Occasionally, the `tns run` will still believe that all it needs to do is sync, whereas the reality is that it should have to rebuild the app. In this case, you will want to use the handy-dandy `--clean` option. This is very important for times when the device does not seems to pick up any of your changes. The `tns run <platform> --clean` command will normally force the app to be rebuilt; however, if `--clean` fails to rebuild, then check out the `tns build` command described later in the chapter. There are a couple of other command parameters that aren't used much, but you might need them for a specific situation. The `--justlaunch` will start the app and do nothing else; `--no-watch` will disable live syncing, and finally `--device <device id>` will force the app to be installed only on a specific device. You can view which devices are available for the installation of the app by running `tns devices`.

tns debug <platform>

The next command we will discuss is `tns debug <platform>`; this will allow you to use the debug tools to test your application. This works in a similar way to the `tns run` command; however, instead of it just running your app, it will debug it. The debugger will use the standard Chrome development tools, which enables you to step through the code: break points, call stacks, and console logs. This command will give you a URL, which you can use to open in Chrome. In iOS specifically, you should run `tns debug ios --chrome` to get the URL for chrome-devtools. The following is an example of debugging Android via the Chrome debugger:

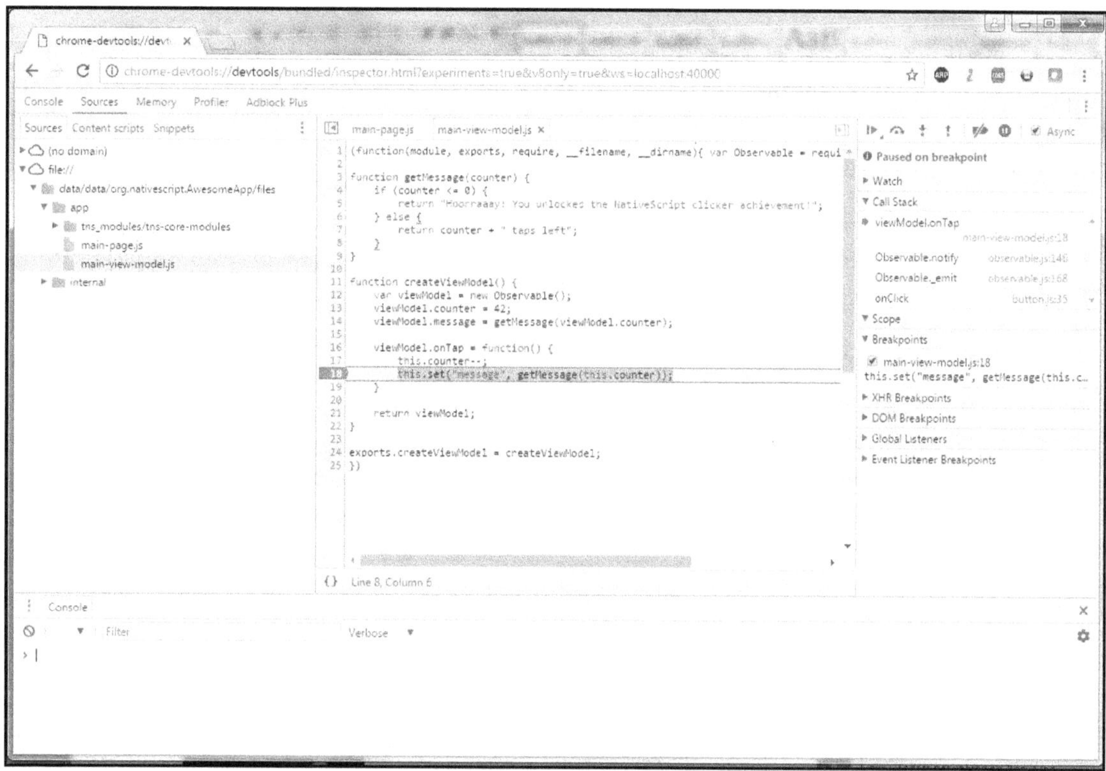

Some of the same `tns run` parameters are valid here, such as `--no-watch`, `--device`, and `--clean`. In addition to those commands, several other commands are available, for example, `--debug-brk`, which is used to make the app break at the start of the application so that you can easily set breakpoints before continuing the start up process. `--start` and `--stop` allow you to attach and detach from an already running application.

Don't forget that if you are currently using the debugger, JavaScript has the cool `debugger;` command, which will force an attached debugger to break just as though you had a breakpoint set. This can be used to put a break anywhere in your code, and it is ignored if a debugger is not attached to your program.

tns build <platform>

The next command you will need to be aware of is `tns build <platform>`; this command fully builds a new app from scratch. Now, the primary use for this command is when you want to build a debug or release version of the app you are going to give to somebody else to test or to upload it to one of the stores. However, it can also be used to force a fully clean build of the application, if the `tns run` version of your app is in a weird state--this will do a full rebuild. If you don't include the `--release` flag, the build will be the default debug build.

On iOS, you will use `--for-device`, which will make the app compile for a real device rather than an emulator. Remember that you need to have signing keys from Apple to do a proper release build.

On Android, when you use `--release`, you will need to include all of the following `--key-store-*` parameters; these are required to sign in to your Android application:

`--key-store-path`	Where your key store file is located.
`--key-store-password`	Your password to read any of the data in your keystore.
`--key-store-alias`	The alias for this app. So, inside your keystore, you may have `AA` as the alias, which in your mind equals AwesomeApp. I prefer to make the alias the same as the full name of the app, but this is your choice.
`--key-store-alias-password`	This is the password required to read the actual signing key assigned to the alias you just set.

Since keystores can be confusing to deal with, we will stray from the path slightly here and discuss how to actually create a keystore. This is normally only a one-time thing, that you will need to do for each Android application you want to release. This is also not something you need to worry about for iOS applications, as Apple provides you the signing keys, and they fully control them.

Android Keystores

On Android, you create your own application signing keys. As such, this key is used for the entire life of your application--by entire, I mean you use the same key to release every version of your application. This key is what links version 1.0 to v1.1 to v2.0. Without using the same key, the application will be considered a totally different application.

The reason there are two passwords is that your keystore can actually contain an unlimited number of keys, and so, each key in the keystore has its own password. Anyone who has access to this key can pretend to be you. This is helpful for building servers, but not so helpful if you lose them. You cannot change the key at a later time, so making backups of your keystore is extremely important.

Without your keystore, you will never be able to release a new version of the exact same application name, meaning that anyone using the older version will not see that you have an updated version. So, again, it is critical that you back up your keystore file.

Creating a new keystore

```
keytool -genkey -v -keystore <keystore_name> -alias <alias_name> keyalg RSA
-keysize 4096 -validity 10000
```

You provide a path to the file you want to save into the `keystore_name`, and for the `alias_name` you put the actual key name for which I normally use the application name; So, you type the following:

```
keytool -genkey -v -keystore android.keystore -alias
com.mastertechapps.awesomeapp -keyalg RSA -keysize 4096 -validity 10000
```

Then, you will see the following:

```
Enter keystore password:
 Re-enter new password:
 What is your first and last name?
   [Unknown]:  Nathanael Anderson
What is the name of your organizational unit?
   [Unknown]:  Mobile Applications
What is the name of your organization?
   [Unknown]:  Master Technology
What is the name of your City or Locality?
   [Unknown]:  Somewhere
What is the name of your State or Province?
   [Unknown]:  WorldWide
```

```
What is the two-letter country code for this unit?
   [Unknown]:  WW
Is CN=Nathanael Anderson, OU=Mobile Applications, O=Master Technology,
L=Somewhere, ST=WorldWide, C=WW correct?
   [no]:  yes
Generating 4,096 bit RSA key pair and self-signed certificate
(SHA256withRSA) with a validity of 10,000 days        for: CN=Nathanael
Anderson, OU=Mobile Applications, O=Master Technology, L=Somewhere,
ST=WorldWide, C=WW
Enter key password for <com.mastertechapps.awesomeapp>
        (RETURN if same as keystore password):
[Storing android.keystore]
```

You now have a keystore for your application.

Android Google Play Fingerprints

If you use Google Play services, you might have to give them your Android application key fingerprint. To get your key fingerprint, you can use the following commands:

```
keytool -list -v -keystore <keystore_name> -alias <alias_name>  -storepass
<password> -keypass <password>
```

You should see something like this:

```
Alias name: com.mastertechapps.awesomeapp
Creation date: Mar 14, 2017
Entry type: PrivateKeyEntry
Certificate chain length: 1
Certificate[1]:
Owner: CN=Nathanael Anderson, OU=Mobile Applications, O=Master Technology,
L=Somewhere, ST=WorldWide, C=WW
Issuer: CN=Nathanael Anderson, OU=Mobile Applications, O=Master Technology,
L=Somewhere, ST=WorldWide, C=WW

Serial number: 2f886ac2

Valid from: Sun Mar 14 14:14:14 CST 2017 until: Thu Aug 17 14:14:14 CDT
2044

Certificate fingerprints:
         MD5:  FA:9E:65:44:1A:39:D9:65:EC:2D:FB:C6:47:9F:D7:FB
         SHA1: 8E:B1:09:41:E4:17:DC:93:3D:76:91:AE:4D:9F:4C:4C:FC:D3:77:E3
         SHA256:
42:5B:E3:F8:FD:61:C8:6E:CE:14:E8:3E:C2:A2:C7:2D:89:65:96:1A:42:C0:4A:DB:63:
D8:99:DB:7A:5A:EE:73
```

Note that in addition to ensuring that you keep a good back up of your keystores, if you ever sell your application to another vendor, having separate keystores per application makes the transfer a lot easier and safer for you. If you use the same keystore and/or alias, it makes it difficult for you to separate out who gets what. So, for the sake of simplicity, I personally recommend that you have a separate keystore and alias per application. I normally keep the keystore in with the app and under version control. Since both opening and accessing an alias are password protected, you are fine unless you choose your passwords poorly.

Back in command

Now that we've taken took a side trip to deal with Android keystores, we will dive deeper into more of the tns commands that you only use occasionally here and there. The first of these is the tns plugin.

The tns plugin command

This one is actually quite important, but it is only used when you want to deal with plugins. The most common version of this command is just `tns plugin add <name>`. So, for example, if you want to install a plugin called *NativeScript-Dom*, you will perform `tns plugin add nativescript-dom`, and it will automatically install the code for this plugin to be used in your application. To remove this plugin, you would type `tns plugin remove nativescript-dom`. We also have `tns plugin update nativescript-dom` to remove the plugin and download and install the newest version of the plugin. Finally, running `tns plugin` alone will give you a list of plugins and their versions that you have installed:

```
D:\projects\nativescript\AwesomeApp>tns plugin
Dependencies:
┌──────────────────────────────┬─────────┐
│ Plugin                       │ Version │
│ nativescript-dom             │ ^1.0.9  │
│ nativescript-theme-core      │ ~1.0.2  │
│ tns-core-modules             │ 2.5.2   │
└──────────────────────────────┴─────────┘
Dev Dependencies:
┌──────────────────────────────────────┬─────────┐
│ Plugin                               │ Version │
│ babel-traverse                       │ 6.4.5   │
│ babel-types                          │ 6.4.5   │
│ babylon                              │ 6.4.5   │
│ lazy                                 │ 1.0.11  │
│ nativescript-dev-android-snapshot    │ ^0.*.*  │
└──────────────────────────────────────┴─────────┘
```

However, to be honest, if I need this information I'm looking for outdated plugins, and so your better bet is to type `npm outdated` and let `npm` give you the list of outdated plugins and the current versions:

```
D:\projects\nativescript\AwesomeApp>npm outdated
Package           Current  Wanted  Latest  Location
babel-traverse      6.4.5   6.4.5  6.23.1
babel-types         6.4.5   6.4.5  6.23.0
babylon             6.4.5   6.4.5  6.16.1
```

If you have outdated plugins, then you can use the `tns plugin update` command to upgrade them.

The tns install <dev_plugin> command

This one isn't used very much, but it is useful when you need it, as it allows you to install development plugins, such as webpack, typescript, coffee script, or SASS support. So, if you decide that you want to use *webpack*, you can type `tns install webpack`, and it will install the webpack support so that you can webpack your application.

The tns create <project_name> command

This command is what we use to create a new project. This will create a new directory and install all the platform-independent code required to build a new app. The important parameters for this are `--ng`, which tells it to use the Angular template (which is what we are using in this book--without `--ng`, you get the plain JS template) and the `--appid`, which allows you to set your full app name. So, `tns create AwesomeApp --ng --appid com.mastertechapps.awesomeapp` will create a new Angular app in the `AwesomeApp` directory with the app ID, `com.mastertechapps.awesomeapp`.

The tns info command

Another useful command for checking the state of the main NativeScript component is `tns info`; this command will actually check your main NativeScript parts and tell you whether anything is out of date:

```
D:\projects\nativescript\AwesomeApp>tns info
All NativeScript components versions information

 Component          Current version   Latest version   Information
 nativescript       2.5.1             2.5.3            Update available
 tns-core-modules   2.5.2             2.5.2            Up to date
 tns-android        2.5.0             2.5.0            Up to date
 tns-ios                              2.5.0            Not installed
```

As you can see from the preceding example, there is a newer version of the NativeScript command-line and I don't have the `ios` runtime installed.

The tns platform [add|remove|clean|upgrade] <platform> command

You can use the `tns platform [add|remove|clean|upgrade] <platform>` command to install, remove, or update the platform modules just like plugins. These are the `tns-android` and `tns-ios` modules you see listed in the prior `tns info` command. The application actually needs these platform-specific modules to be installed. By default, when you perform a `tns run`, it will automatically install them if they are missing. Occasionally, if the application refuses to build, you can use `tns platform clean <platform>`, and it will automatically uninstall and then re-install the platform which will reset the build process.

Note that when you do a `tns platform clean/remove/update`, these will totally delete the `platforms/<platform>` folder. If you have made any manual changes to the files in this folder (which is not recommended), those changes will be deleted.

The tns test <platform> command

The `tns test <platform>` command allows you to install and/or start the testing framework. We will cover testing in more depth in later chapters, however, for the sake of completeness, we will cover the command in this section. `tns test init` will initialize the testing system; you will do this once per application. It will ask you to select a testing framework and then install your chosen testing framework. `tns test <platform>` will start the testing on that specific platform.

The tns device command

If you specifically need to target a device, using the `tns device` command will give you a list of the devices that are installed and connected to your computer. This will allow you to use the `--device <deviceid>` parameter on the `tns run/debug` commands:

```
bash-3.2$ tns device --available-devices
```

#	Device Name	Platform	Device Identifier	Type	Status
1	iPhone 6	iOS	E60921AB-60FE-49CC-B80E-8A3CC8244337	Emulator	Connected

The tns doctor command

The `tns doctor` command checks your environment for common issues. It will attempt to detect whether everything is installed and configured correctly. It mostly works, but occasionally it will fail and state something is broken even when everything actually works. However, it provides a very good first indication of what might be wrong if your `tns run/build/debug` no longer works.

The tns help command

If you totally forget what we have written here, you can execute `tns help` which will give you an overview of the different commands. Some of the parameters may not be listed but at this point, they do exist. In newer versions, newer parameters and commands may be added to `tns`, and this is the easiest way to find out about them.

If, for some reason, your app does not seem to be updating properly, the easiest way to fix this is to uninstall the app from the device. Then, try and do a `tns build <platform>`, then `tns run <platform>`. If that fails to fix it, then uninstall the app again, do a `tns platform clean <platform>`, and then do your `tns run`. Occasionally, the platform may get in a weird state, and resetting it is the only way to fix the problem.

TNS command-line cheatsheet

Command-line	Description
`tns --version`	This returns the version of the NativeScript command. If you are running an older version, then you can use npm to upgrade your NativeScript command like this: `npm install -g nativescript`.
`tns create <your project name>`	This creates a brand new project.The following are its parameters: `--ng` and `--appid`.
`tns platform add <platform>`	This adds a target platform to your project.
`tns platform clean <platform>`	This command is normally not needed, but if you are messing with the platform directory and your platform, you can remove and then add it back. Note that this deletes the entire platform directory. So, if you have made any specific customizations to your Android manifest or iOS Xcode project file, you should back them up before running the clean command.
`tns platform update <platform>`	This is actually a pretty important command. NativeScript is still a very active project that is under a lot of development. This command upgrades your platform code to the latest version, which typically eliminates bugs and adds lots of new features. Note that this should be done alongside an upgrade of the common JavaScript libraries, as most of the time they are in sync with each other.
`tns build <platform>`	This builds the application for that platform using the parameters: `--release`, `--for-device`, and `--key-store-*`.
`tns deploy <platform>`	This builds and deploys the application to a physical or virtual device for that platform.
`tns run <platform>`	This builds, deploys, and starts the application on a physical device or an emulator. This is the command that you will use the majority of the time to run your application and check out the changes. Its parameters are `--clean`, `--no-watch`, and `--justlaunch`.

`tns debug <platform>`	This builds, deploys and then starts the application on a physical device or an emulator in debug mode. This is probably the second most used command. Its parameters are `--clean`, `--no-watch`, `--dbg-break`, and `--start`.
`tns plugin add <plugin>`	This allows you to add a third-party plugin or component. These plugins can be entirely JavaScript-based code, or they might also contain a compilation from the Java or Objective-C library.
`tns doctor`	This allows you to run diagnostic checks on your environment if NativeScript does not appear to be working.
`tns devices`	This shows a list of connected devices for use with the `--device` command.
`tns install <dev plugin>`	This will install a development plugin (that is, webpack, typescript, and so on).
`tns test [ init | <platform> ]`	This allows you to create or run any tests for your application. Using init will initialize the test framework for the application. Then, you can type the platform to run the tests on that platform.

Summary

Now that you have an idea of the power of a command line, all you really need to remember is `tns debug ios` and `tns run android`; these will be your constant friends in our adventure. Throw in a couple of `tns plugin add` commands and then wrap up the application when finally finished with a `tns build`, and you are golden. However, don't forget about the rest of the commands; they all serve a purpose. Some of them are rarely used, but some of them are extremely helpful when you need them.

In `Chapter 7`, *Building the multi-track Player*, we will start exploring how to actually access the Native platform and integrate with plugins.

7
Building the Multitrack Player

We've made it to the keystone of NativeScript development: Direct access to Objective-C/Swift APIs on iOS and Java APIs on Android via TypeScript.

This is, by far, one of the most unique aspects of NativeScript and opens up many opportunities to you as a mobile developer. In particular, our app is going to need to take advantage of rich native audio APIs on both iOS and Android to achieve its core competency of delivering a compelling multitrack recording/mixing experience to our users.

Understanding how to code against these APIs will be essential to unlocking your mobile app's full potential. Additionally, learning how to integrate existing NativeScript plugins, which may already provide consistent APIs on both iOS and Android, can help you reach your goals even faster. Leveraging the best performance each platform can deliver will be the focus of our journey in Part 3.

In this chapter, we will cover the following:

- Integrating the Nativescript-audio plugin
- Creating a model for our track player for future scalability
- Working with RxJS observables
- Understanding Angular's NgZone with third-party libraries and view bindings
- Handling audio playback sync with multiple audio sources

- Taking advantage of Angular's bindings, as well as NativeScript's native event bindings, to achieve the exact usability we're after
- Building a custom shuttle slider for our player controls using Angular platform-specific directives

Implementing our multitrack player via the nativescript-audio plugin

Luckily, the NativeScript community has published a plugin that provides us with a consistent API to use across both iOS and Android to get going with an audio player. Feel free to browse `http://plugins.nativescript.org`, *the official source for NativeScript plugins,* before implementing features, in order to determine if an existing plugin may work for your project.

In this case, the **nativescript-audio** plugin found at `http://plugins.nativescript.org/plugin/nativescript- audio` contains what we need to start integrating the player portion of our app's features, and it works on both iOS and Android. *It even provides a recorder we may be able to use.* Let's start by installing it:

```
npm install nativescript-audio --save
```

The NativeScript framework allows you to integrate with any npm module, opening up a dizzying array of integration possibilities, including NativeScript specific plugins. In fact, if you ever run into a situation where an npm module is giving you trouble (perhaps, because it relies on a node API not compatible in the NativeScript environment), there's even a plugin to help you deal with that at `https://www.npmjs.com/package/nativescript-nodeify`. It is described in detail at `https://www.nativescript.org/blog/how-to-use-any-npm-module-with-nativescript`.

Whenever integrating with a NativeScript plugin, create a model or Angular service around its integration to provide isolation around that integration point.

Try to isolate third-party plugin integration points by creating a reusable model or Angular service around that plugin. This will not only provide your app with nice scalability into the future, but will give you more flexibility down the road if you need to swap that plugin out with something different and/or provide different implementations on iOS or Android.

Building the TrackPlayerModel for our multitrack player

We need each track to have its own instance of an audio player, as well as to expose an API to load the track's audio file. This will also provide a good place to expose the track's duration once the audio file is loaded.

Since this model will likely be shared across the entire app (foreseeably with recording playback in the future as well), we will create this with our other models in `app/modules/shared/models/track-player.model.ts`:

```
// libs
import { TNSPlayer } from 'nativescript-audio';

// app
import { ITrack } from

'./track.model';

interface ITrackPlayer {
  trackId: number;
  duration: number;
  readonly

player: TNSPlayer;
}

export class TrackPlayerModel implements ITrackPlayer {
  public trackId:

number;
  public duration: number;

  private _player: TNSPlayer;

  constructor() {

this._player = new TNSPlayer();
   }

  public load(track: ITrack): Promise<number> {
    return

new Promise((resolve, reject) => {
      this.trackId = track.id;
```

```
        this._player.initFromFile({
          audioFile: track.filepath,
          loop: false
        }).then(() => {


this._player.getAudioTrackDuration()
            .then((duration) => {
              this.duration = +duration;
              resolve();
            });
        });
      });
    }

    public get player():

TNSPlayer {
      return this._player;
    }
}
```

We start by importing the sweet NativeScript community audio player `TNSPlayer` from the `nativescript- audio` plugin. We then define a simple interface to implement for our model which, will reference `trackId`, its `duration`, and a `readonly` getter for the `player` instance. Then, we include that interface to use with our implementation, which constructs an instance of `TNSPlayer` with itself. Since we want a flexible model that can load its track file at any time, we provide a `load` method taking `ITrack` that utilizes the `initFromFile` method. This, in turn, asynchronously fetches the track's total duration (returned as a string, so we use `+duration`) to store the number on the model before resolving the track's initialization completes.

For consistency and standards, just be sure to also export this new model from `app/modules/shared/models/index.ts`:

```
export * from './composition.model';
export * from './track-player.model';
export * from

'./track.model';
```

Lastly, we provide a getter for the player instance that `PlayerService` will use. This brings us to our next step: open `app/modules/player/services/player.service.ts`. We are going to change up our initial implementation a bit with our latest developments; have a look at this in totality and we will explain afterward:

```
// angular
import { Injectable } from '@angular/core';

// libs
import { Subject }

from 'rxjs/Subject';
import { Observable } from 'rxjs/Observable';

// app
import { ITrack, CompositionModel, TrackPlayerModel } from
'../../shared/models';

@Injectable()
export class PlayerService {

  // observable state
  public playing$:

Subject<boolean> = new Subject();
  public duration$: Subject<number> = new Subject

();
  public currentTime$: Observable<number>;

  // active composition
  private _composition: CompositionModel;
  // internal state
  private _playing:

boolean;
  // collection of track players
  private _trackPlayers: Array<TrackPlayerModel>

= [];
  // used to report currentTime from
  private _longestTrack:

TrackPlayerModel;

  constructor() {
    // observe currentTime changes every 1 seconds
```

```
this.currentTime$ = Observable.interval(1000)
      .map(_ =>

this._longestTrack ?
          this._longestTrack.player.currentTime
          : 0);
  }

  public set playing(value: boolean)

{
    this._playing = value;
    this.playing$.next(value);
  }

  public get playing(): boolean {
    return

this._playing;
  }

  public get composition(): CompositionModel

{
    return this._composition;
  }

  public set

composition(comp: CompositionModel) {
    this._composition = comp;


// clear any previous players
    this._resetTrackPlayers();
    // setup

player instances for each track
    let initTrackPlayer = (index: number) => {
      let track = this._composition.tracks[index];
      let trackPlayer = new

TrackPlayerModel();
      trackPlayer.load(track).then(_ => {


this._trackPlayers.push(trackPlayer);
```

```
        index++;
        if (index <

this._composition.tracks.length) {
          initTrackPlayer(index);
        }

else {
          // report total duration of composition

this._updateTotalDuration();
        }
      });

};
    // kick off multi-track player initialization
    initTrackPlayer

(0);
  }

  public togglePlay() {
    this.playing =

!this.playing;
    if (this.playing) {
      this.play();
    } else {
      this.pause();
    }
  }



  public play() {
    for (let t of this._trackPlayers) {
      t.player.play();
    }
  }

  public

pause() {
    for (let t of this._trackPlayers) {
      t.player.pause

();
    }
```

```
  }

  ...

  private

_updateTotalDuration() {
    // report longest track as the total duration of the mix
    let totalDuration = Math.max(
      ...this._trackPlayers.map(t =>

t.duration));
    // update trackPlayer to reflect longest track
    for (let

t of this._trackPlayers) {
      if (t.duration === totalDuration) {

this._longestTrack = t;
        break;
      }
    }


    this.duration$.next(totalDuration);
  }


private _resetTrackPlayers() {
    for (let t of this._trackPlayers) {

t.cleanup();
    }
    this._trackPlayers = [];
  }


}
```

The cornerstone of `PlayerService` at this point is to not only manage the hard work of playing multiple tracks in the mix, but to provide a state, which our views can observe to reflect the composition's state. Hence, we have the following:

```
...
// observable state
public playing$: Subject<boolean> = new Subject();
public duration$:
```

```
Subject<number> = new Subject();
public currentTime$: Observable<number>;

// active

composition
private _composition: CompositionModel;
// internal state
private _playing: boolean;
//

collection of track players
private _trackPlayers: Array<TrackPlayerModel> = [];
// used to report

currentTime from
private _longestTrack: TrackPlayerModel;

constructor() {
  // observe currentTime

changes every 1 seconds
  this.currentTime$ = Observable.interval(1000)
    .map(_ => this._longestTrack ?
      this._longestTrack.player.currentTime
      : 0);
  }
  ...
```

Our view will need to know the playing state as well as `duration` and `currentTime`. Using `Subject` for the `playing$` and `duration$` states will work well, since they are as follows:

- They can emit values directly
- They don't need to emit an initial value
- They don't need any observable composition

On the other hand, `currentTime$` is going to be set up with some composition in mind, since its value will be dependent on an intermittent state that may develop over time (more on this shortly!). In other words, the `playing$` state is a value we control and emit directly via play actions made by the user (or internally based on player state) and the `duration$` state is a value we emit directly as a result of all our track's players becoming initialized and ready.

`currentTime` is a value that the player does not emit automatically via a player event but rather a value we must check for intermittently. Therefore, we compose `Observable.interval(1000)` that will auto emit our mapped value representing the longest track's player instance's actual `currentTime` every 1 second upon its subscription.

The other `private` references help maintain the internal state for the service's use. Most interestingly, we will keep a reference to `_longestTrack`, since our composition's total duration will always be based on the longest track and, hence, will also be used to track `currentTime`.

This set up will provide the essentials of what our view will need to suffice proper user interaction.

RxJS does not include any operators by default. Therefore, `Observable.interval(1000)` *and* `.map` *will crash your app right now if you are to run it!*

The minute you start working more with RxJS, it's a good idea to create an `operators.ts` file to import all your RxJS operators into. Then, import that file in your root `AppComponent`, so you don't end up with those operator imports scattered everywhere throughout your codebase.

Create `app/operators.ts` with the following:

```
import 'rxjs/add/operator/map';
import 'rxjs/add/observable/interval';
```

Then, open `app/app.component.ts` and import that file on the very first line:

```
import './operators';
...
```

Now, we are free to use map, interval, and any other `rxjs` operators we need anywhere in our code, provided we import them into that single file.

The next bit of our service is rather self-explanatory:

```
public set playing(value: boolean) {
  this._playing = value;
  this.playing$.next(value);
}

public get playing(): boolean {
  return this._playing;
}
```

```
public get composition(): CompositionModel

{
  return this._composition;
}
```

Our `playing` setter ensures that the internal state, `_playing`, is updated, as well as our `playing$` subject's value emitted for any subscribers needing to react to this state change. Convenient getters are also added for good measure. The next setter for our composition gets rather interesting, as this is where we interact with our new `TrackPlayerModel`:

```
public set composition(comp: CompositionModel) {
  this._composition = comp;

  // clear any previous

players
  this._resetTrackPlayers();
  // setup player instances for each track
  let initTrackPlayer =

(index: number) => {
    let track = this._composition.tracks[index];
    let trackPlayer = new

TrackPlayerModel();
    trackPlayer.load(track).then(_ => {

      this._trackPlayers.push

(trackPlayer);
      index++;
      if (index < this._composition.tracks.length) {

initTrackPlayer(index);
      } else {
        // report total duration of composition

this._updateTotalDuration();
      }
    });
  };
  // kick off multi-track player initialization


 initTrackPlayer(0);
}
...
```

```
private _resetTrackPlayers() {
  for (let t of this._trackPlayers) {

 t.cleanup();
  }
  this._trackPlayers = [];
}
```

Whenever we set the active composition, we first ensure our service's internal `_trackPlayers` reference is properly cleaned up and cleared with `this._resetTrackPlayers()`. We then set up a local method `initTrackPlayer` that can be called iteratively, given the async nature of each player's `load` method to ensure each track's player is properly loaded with the audio file, including its duration. After each successful load, we add to our collection of `_trackPlayers`, iterate, and continue until all the audio files are loaded. When complete, we call `this._updateTotalDuration()` to determine the final duration of our composition of tracks:

```
private _updateTotalDuration() {
  // report longest track as the total duration of the mix
  let

totalDuration = Math.max(
    ...this._trackPlayers.map(t => t.duration));
  // update trackPlayer to reflect

longest track
  for (let t of this._trackPlayers) {
    if (t.duration === totalDuration) {

this._longestTrack = t;
      break;
    }
  }
  this.duration$.next(totalDuration);
}
```

Since the track with the longest duration should always be used to determine the total duration of the entire composition, we use `Math.max` to determine what the longest duration is and then store a reference to the track. Because multiple tracks could have the same duration, it doesn't really matter which track is used, just as long as one matches the longest duration. This `_longestTrack` will be our *pace setter* if you will, as it will be used to determine `currentTime` of the entire composition. Lastly, we emit the longest duration as `totalDuration` via our `duration$` subject for any subscribing observers.

The next couple of methods provide the basis of our composition's overall playback control:

```
public togglePlay() {
  this.playing = !this.playing;
  if (this.playing) {
    this.play();
  }

else {
    this.pause();
  }
}

public play() {
  for (let t of this._trackPlayers) {

 t.player.play();
  }
}

public pause() {
  for (let t of this._trackPlayers) {

t.player.pause();
  }
}
```

Our primary play button in our UI will use the `togglePlay` method to control playback and, hence, is used to toggle the internal state as well as engage all the track player's play or pause methods.

Let the music play!

To try all this out, let's add three sample audio files from a jazz track composed by the exquisite *Jesper Buhl Trio* called *What Is This Thing Called Love*. The tracks are already separated by drums, bass, and piano. We can add these `.mp3` files to an `app/audio` folder.

Let's modify our demo composition's tracks in `MixerService` to provide references to these new real audio files. Open `app/modules/mixer/services/mixer.service.ts` and make the following modifications:

```
private _demoComposition(): Array<IComposition> {
  // starter composition for user to demo on first

launch
```

```
  return [
    {
      id: 1,
      name: 'Demo',
      created: Date.now(),

order: 0,
      tracks: [
        {
          id: 1,
          name: 'Drums',
          order: 0,
          filepath:

'~/audio/drums.mp3'
        },
        {

id: 2,
          name: 'Bass',
          order: 1,

    filepath: '~/audio/bass.mp3'
        },
        {
          id: 3,
          name: 'Piano',
          order:

2,
          filepath: '~/audio/piano.mp3'
        }
      ]
    }
  ];
}
```

Let's now provide an input to our player controls, which will take our selected composition. Open `app/modules/mixer/components/mixer.component.html`, and make the following highlighted modification:

```
<action-bar [title]="composition.name"></action-bar>
<GridLayout rows="*, auto" columns="*"

class="page">
  <track-list [tracks]="composition.tracks" row="0" col="0">
  </track-list>
```

```
<player-controls [composition]="composition"
    row="1" col="0"></player-controls>
</GridLayout>
```

Then, in `PlayerControlsComponent` at `app/modules/player/components/player-controls/player-controls.component.ts`, we can now observe the state of `PlayerService` via its various observables:

```
// angular
import { Component, Input } from '@angular/core';

// libs
import { Subscription } from 'rxjs/Subscription';

// app
import { ITrack,

CompositionModel } from '../../../shared/models';
import { PlayerService } from '../../services';

@Component({
  moduleId: module.id,
  selector: 'player-controls',
  templateUrl: 'player-

controls.component.html'
})
export class PlayerControlsComponent {

  @Input() composition:

CompositionModel;

  // ui state
  public playStatus: string = 'Play';
  public duration:

number = 0;
  public currentTime: number = 0;

  // manage subscriptions
  private _subPlaying:

Subscription;
  private _subDuration: Subscription;
  private _subCurrentTime:

Subscription;
```

```
  constructor(
    private playerService: PlayerService
  ) { }


public togglePlay() {
    this.playerService.togglePlay();
  }



  ngOnInit() {
    // init audio player for composition

this.playerService.composition = this.composition;
    // react to play state

this._subPlaying = this.playerService.playing$
      .subscribe((playing: boolean) =>

{
        // update button state
        this._updateStatus(playing);
        //

update slider state
        if (playing) {
          this._subCurrentTime =

this.playerService
              .currentTime$
              .subscribe

((currentTime: number) => {
                this.currentTime = currentTime;

           });
        } else if (this._subCurrentTime) {

this._subCurrentTime.unsubscribe();
        }
      });
    //

update duration state for slider
    this._subDuration = this.playerService.duration$

   .subscribe((duration: number) => {
```

```
        this.duration = duration;

   });
    }

    ngOnDestroy() {
      // cleanup

  if (this._subPlaying)
        this._subPlaying.unsubscribe();
      if

  (this._subDuration)
        this._subDuration.unsubscribe();
      if

  (this._subCurrentTime)
        this._subCurrentTime.unsubscribe();
    }



    private _updateStatus(playing: boolean) {
      this.playStatus =

  playing ? 'Stop' : 'Play';
    }
  }
```

The cornerstone of `PlayerControlComponent` is now its ability to set the active composition via `this.playerService.composition = this.composition` inside `ngOnInit`, which is when the composition input is ready, as well as subscribe to the various states provided by `PlayerService` to update our UI. Most interesting here is the `playing$` subscription that manages the `currentTime$` subscription based on whether it's playing or not. If you recall, our `currentTime$` observable started with `Observable.interval(1000)`, meaning every one second, it will emit the longest track's `currentTime`, shown here again for reference:

```
  this.currentTime$ = Observable.interval(1000)
    .map(_ => this._longestTrack ?

  this._longestTrack.player.currentTime
      : 0);
```

We only want to update `currentTime` of `Slider` when playback is engaged; hence, the subscription when the `playing$` subject emit is `true`, which will allow our component to receive the player's `currentTime` every second. When `playing$` emit is `false`, we unsubscribe, to no longer receive the `currentTime` updates. Excellent.

We also subscribe to our `duration$` subject to update the Slider's maxValue. Lastly, we ensure all subscriptions are cleaned up via their `Subscription` references inside `ngOnDestroy`.

Let's take a look at our view bindings now for `PlayerControlsComponent` at `app/modules/player/components/player-controls/player-controls.component.html`:

```
<GridLayout rows="100" columns="100,*"
  row="1" col="0" class="p-x-10">
  <Button [text]

="playStatus" (tap)="togglePlay()"
    row="0" col="0" class="btn btn-primary w-

100"></Button>
  <Slider [maxValue]="duration" [value]="currentTime"
    minValue="0" row="0" col="1" class="slider">
  </Slider>
</GridLayout>
```

If you run the app, you can now select the **Demo** composition and play music on both iOS and Android.

MUSIC TO OUR EARS! This is pretty awesome. In fact, it's friggin' sweet!!

There are a couple things you may notice or desire at this point:

- After choosing the Play button, it properly changes to Stop, but when playback reaches the end, it does not return to its original Play text.
- `Slider` should also return to position 0 to reset playback.
- The total `duration` and `currentTime` on iOS uses seconds; however, Android uses milliseconds.
- On iOS, you may notice a very subtle playback sync issue on all the tracks if you choose to play/pause many times during the playback of the composition's demo tracks.

- The current time and duration labels are needed.
- **Playback seeking** would be nice to be able to shuttle our slider to control the position of playback.

Polishing the implementation

We're missing a few important pieces in our model and service to really polish off our implementation. Let's start with handling completion and error conditions with our track player instances. Open `TrackPlayerModel` at `app/modules/shared/models/track-player.model.ts` and add the following:

```
...
export interface IPlayerError {
  trackId: number;

error: any;
}

export class TrackPlayerModel implements ITrackPlayer {
  ...
  private _completeHandler: (number) => void;
  private _errorHandler:

(IPlayerError) => void;

  ...

  public load(
    track: ITrack,

complete: (number) => void,
    error: (IPlayerError) => void
  ):

Promise<number> {
    return new Promise((resolve, reject) => {
      ...

this._completeHandler = complete;
      this._errorHandler = error;


this._player.initFromFile({
        audioFile: track.filepath,
        loop: false,
```

```
completeCallback: this._trackComplete.bind(this),
        errorCallback:

this._trackError.bind(this)
  ...

  private _trackComplete(args: any) {
    // TODO:

works well for multi-tracks with same length
    // may need to change in future with varied lengths

this.player.seekTo(0);
    console.log('trackComplete:', this.trackId);
    if (this._completeHandler)

this._completeHandler(this.trackId);
  }

  private _trackError(args: any) {
    let error =

args.error;
    console.log('trackError:', error);
    if (this._errorHandler)
      this._errorHandler({

trackId: this.trackId, error });
 }
```

We start by defining the shape of each track error with `IPlayerError`. Then, we define references to the `_completeHandler` and `_errorHandler` functions captured via the `load` arguments, which now require complete and error callbacks. We assign those both before assigning the model's internal `this._trackComplete` and `this._trackError` (*using the* `.bind(this)` *syntax to ensure the function scope is locked to itself*) to `completeCallback` and `errorCallback` of `TNSPlayer`.

`completeCallback` and `errorCallback` will fire outside the zone. This is why we inject `NgZone` and use `ngZone.run()` later in the chapter. We can avoid that by creating a callback with the `zonedCallback` function. It will make sure that the callback will be executed in the same zone as the code that creates the callback. For example:

```
this._player.initFromFile({
  audioFile: track.filepath,
  loop: false,
```

```
  completeCallback:

zonedCallback(this._trackComplete.bind(this)),
  errorCallback:

zonedCallback(this._trackError.bind(this))
  ...
```

This provides us the ability to internally handle each condition before dispatching out those conditions.

One such internal condition is resetting each audio player back to zero when it completes playing, so we simply call the `seekTo` method of `TNSPlayer` to reset it. We mark a *TODO,* because although this works well when all the tracks are the same length (*as is the case with our Demo tracks*), this will most certainly become potentially problematic in the future when we start recording our own varied multitracks with different lengths. Imagine we have two tracks in a composition: track 1 with a duration of 1 minute and track 2 a duration of 30 seconds. If we play the composition to 45 seconds and hit pause, track 2 would have called its completion handler already and reset back to 0. We then hit play to resume. Track 1 resumes from 45 seconds but track 2 is back at 0. *We will address that when we get there, so don't fret about it!* At this point, we are polishing our first phase implementation.

Lastly, we call out to the assigned `completeHandler` to let the caller know which trackId has completed. For `trackError`, we simply call out passing along `trackId` and `error`.

Now, let's go back to `PlayerService` and wire this in. Open `app/modules/player/services/player.service.ts` and make the following modifications:

```
// app
import { ITrack, CompositionModel, TrackPlayerModel, IPlayerError } from

'../../shared/models';

@Injectable()
export class PlayerService {

  // observable state
  ...
  public complete$: Subject<number> = new Subject();
  ...
  public set

composition(comp: CompositionModel) {
    ...
    let initTrackPlayer = (index:
```

```
number) => {
      ...
      trackPlayer.load(
        track,

   this._trackComplete.bind(this),
        this._trackError.bind(this)

  ...


 private _trackComplete(trackId: number) {
    console.log('track complete:', trackId);
    this.playing =

false;
    this.complete$.next(trackId);
  }

  private _trackError(playerError: IPlayerError) {

  console.log(`trackId ${playerError.trackId} error:`,
      playerError.error);
  }
  ...
```

We've added another subject, `complete$`, to allow view components to subscribe to when the track playback completes. Additionally, we have added two callback handlers, `_trackComplete` and `_trackError`, which we pass along to our `load` method of `TrackPlayerModel`.

However, if we were to try and update view bindings as a result of the `complete$` subscriptions firing in any view component, you would notice something puzzling. **The view would not update!**

Anytime you integrate with third-party libraries, take note of callback handlers coming from the library, which you may intend to update a view binding. Inject NgZone and wrap with `this.ngZone.run(() => ...` where needed.

Third-party libraries that provide callbacks may often need to run through Angular's NgZone. The great folks at Thoughtram published a great article on Zones if you'd like to learn more, at `https://blog.thoughtram.io/angular/2016/02/01/zones-in-angular-2.html`.

The third-party library **nativescript-audio** integrates with the iOS and Android native audio players and provides callbacks you can wire up to handle completion and error conditions. These callbacks are executed asynchronously within the context of the native audio players and, because they are not handled within the context of user events like a tap, or a result of a network request, or a timer like `setTimeout`, we need to ensure the result and the subsequent code execution take place within Angular's NgZone if we intend them to result in updating view bindings.

Since we intend for the `complete$` subject to result in view binding updates (*specifically, resetting our slider*), we will inject NgZone and wrap our callback handling. Back in `app/modules/player/services/player.service.ts`, let's make the following adjustment:

```
// angular
import { Injectable, NgZone } from '@angular/core';

@Injectable()


export class PlayerService {

  ...
  constructor(private ngZone: NgZone) {}

...
  private _trackComplete(trackId: number) {
    console.log('track complete:', trackId);

this.ngZone.run(() => {
      this.playing = false;
      this.complete$.next(trackId);


    });
  }
  ...
```

Now, we will be clear when using this new `complete$` subject to react to our service's state in our view components. Let's adjust `PlayerControlsComponent` at `app/modules/player/components/player- controls/player-controls.component.ts` to observe the `complete$` subject to reset our `currentTime` binding:

```
export class PlayerControlsComponent {

  ...
```

```
  private _subComplete: Subscription;
  ...
  ngOnInit() {
    ...
    // completion should reset currentTime
    this._subComplete

= this.playerService.complete$.subscribe(_ => {
      this.currentTime = 0;


    });
  }
  ngOnDestroy() {
    ...
    if (this._subComplete)

this._subComplete.unsubscribe();
  }
  ...
```

iOS Audio Player reports `duration` and `currentTime` in seconds, whereas Android reports in milliseconds. We need to standardize that!

Let's add a method to `PlayerService` to standardize the time, so we can rely on both the platforms providing time in seconds:

```
...
// nativescript
import { isIOS } from 'platform';
...

@Injectable()
export class PlayerService {

 constructor() {
   // observe currentTime changes

every 1 seconds
   this.currentTime$ = Observable.interval(1000)
     .map(_ => this._longestTrack ?

  this._standardizeTime(
         this._longestTrack.player.currentTime)
```

```
: 0;
      );
  }
  ...
  private _updateTotalDuration() {
    ...
    // iOS: reports

duration in seconds
    // Android: reports duration in milliseconds
    //

standardize to seconds
    totalDuration = this._standardizeTime(totalDuration);

console.log('totalDuration of mix:', totalDuration);
    this.duration$.next(totalDuration);
  }
  ...


private _standardizeTime(time: number) {
    return isIOS ? time : time * .001;
  }
  ...
```

We are able to take advantage of the `isIOS` Boolean provided by the `platform` module from NativeScript to conditionally adjust our time for Android's milliseconds to seconds conversion.

Using the `isIOS` and/or `isAndroid` Boolean from NativeScript's `platform` module is a very effective way to make platform adjustments across your codebase where needed.

So what about that subtle playback sync issue with multiple tracks on iOS ?

On iOS, you may notice a very subtle playback sync issue on all the tracks if you choose play/pause many times during the 14 seconds of playback on the composition's demo tracks. We could surmise this could also happen on Android at some point.

Using NativeScript's strengths by tapping directly into the native API of the underlying iOS AVAudioPlayer instance from the nativescript-audio plugin

Let's insert some safeguards into our play/pause logic to help ensure our tracks stay in sync to the best of our programming abilities. The **nativescript-audio** plugin offers an iOS-only method called `playAtTime`. It works in tandem with the special `deviceCurrentTime` property, as described in Apple's documentation for this very purpose at `https://developer.apple.com/reference/avfoundation/avaudioplayer/1387462-devicecurrenttime? language=objc`.

Since `deviceCurrentTime` is not exposed by the nativescript-audio plugin, we can access the native property directly via the `ios` getter. Let's adjust the `play` method of `PlayerService` to use it:

```
public play() {
  // for iOS playback sync
  let shortStartDelay = .01;
  let

now = 0;

  for (let i = 0; i < this._trackPlayers.length; i++) {

let track = this._trackPlayers[i];
    if (isIOS) {
      if (i == 0) now =

track.player.ios.deviceCurrentTime;
      (<any>track.player).playAtTime

(now + shortStartDelay);
    } else {
      track.player.play

();
    }
  }
}
```

Since `track.player` is our instance of `TNSPlayer`, we can access the underlying native platform player instance (for iOS, it's `AVAudioPlayer`) via its **ios** getter to access `deviceCurrentTime` directly. We provide a very short start delay for good measure, add that into the first track's `deviceCurrentTime`, and use that to start all of our tracks at precisely the same time, which works wonderfully! Because `playAtTime` is not published via the TypeScript definitions with the nativescript-audio plugin, we simply type-cast the player instance (`<any>track.player`) before calling the method to suffice the tsc compiler. Since there is no equivalent on Android, we will just use the standard media player's play method, which works well for Android.

Let's now adjust our pause method with a similar safeguard:

```
public pause() {
  let currentTime = 0;

  for (let i = 0; i <

this._trackPlayers.length; i++) {
    let track = this._trackPlayers[i];
    if

(i == 0) currentTime = track.player.currentTime;
    track.player.pause();
    // ensure tracks pause

and remain paused at the same time
    track.player.seekTo(currentTime);
  }
}
```

By using the first track's `currentTime` as the **pace setter**, we pause each track in our mix and ensure they remain at exactly the same time by seeking to the same `currentTime` immediately after pausing. This helps ensure that, when we resume play, they all start from the same point in time. Let's put all this to use in the next section when we build a custom shuttle slider.

Creating a custom ShuttleSliderComponent

We can't have a multitrack studio experience without the ability to shuttle back and forth through our mix! Let's double down on `Slider` and enhance its capabilities by combining the best of all the options NativeScript and Angular provide us. In the process, our player controls will start to become much more useful.

Starting at the high level, open `app/modules/player/components/player-controls/player-controls.component.html` and replace it with the following:

```
<StackLayout row="1" col="0" class="controls">
  <shuttle-slider [currentTime]

="currentTime"
    [duration]="duration"></shuttle-slider>
  <Button

[text]="playStatus" (tap)="togglePlay()"
    class="btn btn-primary w-100"></Button>
</StackLayout>
```

We are replacing `GridLayout` with `StackLayout` to change up our player control's layout a bit. Let's go with a full-width slider stacked on top of our play/pause button. What we're after is similar to the Apple Music app on an iPhone, where the slider is full width with the current time and duration displayed underneath. Now, let's build our custom `shuttle-slider` component and create `app/modules/player/components/player-controls/shuttle- slider.component.html` with the following:

```
<GridLayout #sliderArea rows="auto, auto" columns="auto,*,auto"
  class="slider-area">
  <Slider

#slider slim-slider minValue="0" [maxValue]="duration"
      colSpan="3" class="slider"></Slider>

<Label #currentTimeDisplay text="00:00" class="h4 m-x-5" row="1" col="0">
  </Label>
  <Label

[text]="durationDisplay" class="h4 text-right m-x-5"
    row="1" col="2"></Label>
</GridLayout>
```

Here's where things are going to get very interesting. We are going to combine Angular bindings where useful, like these bindings: `[maxValue]="duration"` and `[text]="durationDisplay"` . However for the rest of our usability wiring we will want more fine grained and manual control. For instance, our containing `GridLayout` via `#sliderArea` is going to be the area the user is going to be able to touch to shuttle back/forth instead of the `Slider` component itself, and we are going to completely disable user interaction with the Slider itself (hence, the `slim-slider` directive attribute, you see). The slider is going to instead be used just for its visual representation of time.

The reason we will be doing this is because we want this interaction to kick off several programmatic actions:

- Pause playback (if playing) while shuttling
- Update the current time display label as we move back/forth
- Kick off the `seekTo` commands to our track player's instances in a controlled manner; hence, reducing extraneous seek commands
- Resume playback when no longer shuttling if it was playing before attempting to shuttle

If we used `Slider` with an Angular binding to `currentTime` via the `currentTime$` observable, which in turn was being controlled by our interaction with it in addition to the state of our track's players, things would be coupled too tightly to achieve the fine grain control we need.

The beauty of what we are about to do here serves as an exemplary testament to how flexible the combination of Angular with NativeScript really is. Let's start programming our interactions in `app/modules/player/components/player-controls/shuttle-slider.component.ts`; here's the complete setup for you to view in full, which we will break down in a moment:

```
// angular
import { Component, Input, ViewChild, ElementRef } from '@angular/core';

//

nativescript
import { GestureTypes } from 'ui/gestures';
import { View } from 'ui/core/view';
import { Label

} from 'ui/label';
import { Slider } from 'ui/slider';
import { Observable } from 'data/observable';
import

{ isIOS, screen } from 'platform';

// app
import { PlayerService } from '../../services';

@Component({
  moduleId: module.id,
  selector: 'shuttle-slider',
  templateUrl: 'shuttle-
```

```
slider.component.html',
  styles: [`
    .slider-area {
      margin: 10 10 0 10;
    }

.slider {
      padding:0;
      margin:0 0 5 0;
      height:5;
    }
  `]
})
export

class ShuttleSliderComponent {

  @Input() currentTime: number;
  @Input() duration: number;


 @ViewChild('sliderArea') sliderArea: ElementRef;
  @ViewChild('slider') slider: ElementRef;

@ViewChild('currentTimeDisplay') currentTimeDisplay: ElementRef;

  public durationDisplay: string;

  private _sliderArea: View;
  private _currentTimeDisplay: Label;
  private _slider: Slider;
  private

_screenWidth: number;
  private _seekDelay: number;

  constructor(private playerService: PlayerService) {

}

  ngOnChanges() {
    if (typeof this.currentTime == 'number')    {
      this._updateSlider

(this.currentTime);
    }
    if (this.duration) {
      this.durationDisplay =
```

```
this._timeDisplay(this.duration);
    }
  }

  ngAfterViewInit() {
    this._screenWidth =

screen.mainScreen.widthDIPs;
    this._sliderArea = <View>this.sliderArea

.nativeElement;
    this._slider = <Slider>this.slider.nativeElement;
    this._currentTimeDisplay =

<Label>this.currentTimeDisplay
                                   .nativeElement;

this._setupEventHandlers();
  }

  private _updateSlider(time: number) {
    if (this._slider)

this._slider.value = time;
    if (this._currentTimeDisplay)
      this._currentTimeDisplay
        .text =

this._timeDisplay(time);
  }

  private _setupEventHandlers() {
    this._sliderArea.on

(GestureTypes.touch, (args: any) => {
      this.playerService.seeking = true;
      let x = args.getX();

      if (x >= 0) {
        let percent = x / this._screenWidth;
        if (percent > .5) {

        percent += .05;
        }
        let seekTo = this.duration * percent;
        this._updateSlider

(seekTo);
```

```
            if (this._seekDelay) clearTimeout(this._seekDelay);
            this._seekDelay = setTimeout

    (() => {
              // android requires milliseconds
              this.playerService
                .seekTo

    (isIOS ? seekTo : (seekTo*1000));
            }, 600);
          }
        });
      }

      private

    _timeDisplay(seconds: number): string {
        let hr: any = Math.floor(seconds / 3600);
        let min: any =

    Math.floor((seconds - (hr * 3600))/60);
        let sec: any = Math.floor(seconds - (hr * 3600)

    - (min * 60));
        if (min < 10) {
          min = '0' + min;
        }
        if (sec < 10){

    sec = '0' + sec;
        }
        return min + ':' + sec;
      }
    }
```

For a rather small component footprint, there's a ton of great stuff going on here! Let's break it down.

Let's look at those property decorators, starting with `@Input`:

```
@Input() currentTime: number;
@Input() duration: number;

// allows these property bindings to flow into our view:
<shuttle-slider
  [currentTime]

="currentTime"
```

```
    [duration]="duration">
  </shuttle-slider>
```

Then, we have our @ViewChild references:

```
  @ViewChild('sliderArea') sliderArea: ElementRef;
  @ViewChild('slider')

  slider: ElementRef;
  @ViewChild('currentTimeDisplay') currentTimeDisplay: ElementRef;

  private _sliderArea: StackLayout;
  private _currentTimeDisplay: Label;
  private _slider: Slider;

  // provides us with references to these view components
  <StackLayout

  #sliderArea class="slider-area">
    <Slider #slider slim-slider

  minValue="0 [maxValue]="duration" class="slider">
    </Slider>
    <GridLayout rows="auto"

  columns="auto,*,auto"
      class="m-x-5">
      <Label #currentTimeDisplay text="00:00"

  class="h4"
        row="0" col="0"></Label>
      <Label [text]="durationDisplay" class="h4 text-right"


        row="0" col="2"></Label>
    </GridLayout>
  </StackLayout>
```

We can then access those ElementRef instances in our component to programmatically work with them; however, not right away. Since ElementRef is a proxy wrapper to the view component, its underlying nativeElement (our actual NativeScript component) is only accessible once Angular's component lifecycle hook ngAfterViewInit fires.

Learn all about Angular's component lifecycle hooks here: https://angular.io/docs/ts/latest/guide/lifecycle- hooks.html.

Therefore, we assign private references to our actual NativeScript components here:

```
ngAfterViewInit() {
  this._screenWidth = screen.mainScreen.widthDIPs;
  this._sliderArea =

<StackLayout>this.sliderArea
                                        .nativeElement;
  this._slider = <Slider>this.slider.nativeElement;
  this._currentTimeDisplay =

<Label>this.currentTimeDisplay

.nativeElement;
  this._setupEventHandlers();
}
```

We also take this opportunity to reference the overall screen width using the **density-independent pixel (dip)** units via the `screen` utility from the `platform` module. This will allow us to do some calculations using our user's finger position on our `sliderArea` StackLayout to adjust the actual value of `Slider`. We then make a call to set up our essential event handlers.

Using our `_sliderArea` reference to the containing StackLayout, we add a `touch` gesture listener to capture any touches the user makes to our slider area:

```
private _setupEventHandlers() {
  this._sliderArea.on(GestureTypes.touch, (args: any) => {

this.playerService.seeking = true; // TODO

    let x = args.getX();
    if (x >= 0) {

  // x percentage of screen left to right
      let percent = x / this._screenWidth;
      if (percent > .5)

{
        percent += .05; // non-precise adjustment
      }
      let seekTo = this.duration * percent;
      this._updateSlider(seekTo);

      if (this._seekDelay) clearTimeout(this._seekDelay);

this._seekDelay = setTimeout(() => {
```

```
        // android requires milliseconds

this.playerService.seekTo(
         isIOS ? seekTo : (seekTo*1000));
      }, 600);
    }
  });
}
```

This allows us to grab the X position of their finger via `args.getX()`. We use that to divide into the user's device screen width to determine a percentage from the left to the right. Since our calculation is not exactly precise, we make a small adjustment when the user passes the 50% mark. This usability works well for our use case right now, but we will reserve the option to improve that later; however, it's perfectly fine for now.

We then multiply the duration by this percentage to get our `seekTo` mark to update our value of `Slider` in order to get immediate UI updates using manual precision:

```
private _updateSlider(time: number) {
  if (this._slider) this._slider.value = time;
  if

(this._currentTimeDisplay)
    this._currentTimeDisplay.text = this._timeDisplay(time);
}
```

Here, we are actually using our NativeScript components directly without Angular's bindings or NgZone in the mix. This can be very handy in cases where you need fine grained and performance control of your UI. Since we want the `Slider` track to move immediately with the user's finger, as well as the time display label formatted with standard musical timecode to represent real time as they interact, we set their values directly at the appropriate time.

We then use a seek delay timeout to ensure we don't make extraneous seek commands to our multitrack player. Each movement by the user will further delay making an actual seek command until they rest where they want it. We also use our `isIOS` Boolean to convert the time as appropriately needed by each platform audio player (seconds for iOS and milliseconds for Android).

Most interesting might be our `ngOnChanges` lifecycle hook:

```
ngOnChanges() {
  if (typeof this.currentTime == 'number') {
    this._updateSlider(this.currentTime);


 }
  if (this.duration) {
    this.durationDisplay = this._timeDisplay(this.duration);
  }
}
```

Angular calls its `ngOnChanges()` method whenever it detects changes to the input properties of the component (or directive).

This is a wonderful way for `ShuttleSliderComponent` to react to its `Input` property changes, `currentTime`, and `duration`. Here, we simply update our slider and the current time display label manually via `this._updateSlider(this.currentTime)` only when it does fire with a valid number. Lastly, we also ensure we update our duration display label. This method will fire every second the PlayerService's `currentTime$` observable fires while an active subscription exists. **Nice!** Oh, and don't forget to add `ShuttleSliderComponent` to the `COMPONENTS` array to be included with the module.

We now need to actually implement this:

```
this.playerService.seeking = true; // TODO
```

We are going to use a couple more nifty observable tricks with our seeking state. Let's open our PlayerService in `app/modules/player/services/player.service.ts` and add the following:

```
...
export class PlayerService {

  ...
  // internal state
  private _playing: boolean;
  private _seeking: boolean;
  private _seekPaused: boolean;

private _seekTimeout: number;
  ...
  constructor(private ngZone: NgZone) {
    this.currentTime$ =
```

```
Observable.interval(1000)
      .switchMap(_ => {
        if (this._seeking)

{
          return Observable.never();
        } else if

(this._longestTrack) {
          return Observable.of(
            this._standardizeTime(

this._longestTrack.player.currentTime));
        } else {
          return Observable.of(0);
        }


   });
  }
  ...
  public set seeking(value: boolean) {
    this._seeking =

value;
    if (this._playing && !this._seekPaused) {
      // pause

while seeking
      this._seekPaused = true;
      this.pause();
    }
    if (this._seekTimeout) clearTimeout(this._seekTimeout);

this._seekTimeout = setTimeout(() => {
      this._seeking = false;
      if

(this._seekPaused) {
        // resume play
        this._seekPaused =

false;
        this.play();
      }
    },

1000);
```

```
    }

    public seekTo(time: number) {
      for

  (let track of this._trackPlayers) {
        track.player.seekTo(time);
      }


    }
    ...
```

We are introducing three new observable operators `switchMap`, `never`, and `of`, which we need to ensure are also imported in our `app/operators.ts` file:

```
import 'rxjs/add/operator/map';
import 'rxjs/add/operator/switchMap';
import

'rxjs/add/observable/interval';
import 'rxjs/add/observable/never';
import

'rxjs/add/observable/of';
```

`switchMap` allows our observable to switch streams based on several conditions, helping us to manage whether `currentTime` needs to emit updates or not. Clearly, when seeking, we don't need to react to the `currentTime` changes. Therefore, we switch our Observable stream to `Observable.never()` while `this._seeking` is true, ensuring our observer is never called.

In our `seeking` setter, we adjust the internal state reference (`this._seeking`) ,and if it was currently `this._playing` and had not yet been paused due to seeking (hence, `!this._seekPaused`), we immediately pause playback (only once). We then set up another timeout to delay resuming playback an additional 400 milliseconds after `seekTo` has been fired from the component if it was playing when seek started (hence, the check on `this._seekPaused`).

This way, the user is free to move their finger across our shuttle slider as much as they'd like and as quickly as they'd like. They will see immediate UI updates to the `Slider` track as well as the current time display label in real time; all the while we are avoiding extraneous `seekTo` commands being sent to our multitrack player until they come to rest, providing a really nice user experience.

Creating SlimSliderDirective for iOS and Android native API modifications

We still have a directive to create for that `slim-slider` attribute we had on `Slider`:

```
<Slider #slider slim-slider minValue="0" [maxValue]="duration"

class="slider"></Slider>
```

We are going to create platform-specific directives, since we will tap into the slider's actual native API on iOS and Android to disable user interaction and hide the thumb for a seamless appearance.

For iOS, create `app/modules/player/directives/slider.directive.ios.ts` with the following:

```
import { Directive, ElementRef } from '@angular/core';

@Directive({
 selector: '[slim-

slider]'
})
export class SlimSliderDirective {

  constructor(private el: ElementRef) { }


ngOnInit() {
    let uiSlider = <UISlider>this.el.nativeElement.ios;
    uiSlider.userInteractionEnabled =

false;
    uiSlider.setThumbImageForState(
      UIImage.new(), UIControlState.Normal);
  }
}
```

We gain access to the underlying native iOS `UISlider` instance via NativeScript's `ios` getter off the `Slider` component itself. We use Apple's API reference documentation (`https://developer.apple.com/reference/uikit/uislider`) to locate an appropriate API to disable interaction via the `userInteractionEnabled` flag and hide the thumb by setting a blank as the thumb. Perfect.

For Android,
create `app/modules/player/directives/slider.directive.android.ts` with the following:

```
import { Directive, ElementRef } from '@angular/core';

@Directive({
  selector: '[slim-

slider]'
})
export class SlimSliderDirective {

  constructor(private el: ElementRef) { }


ngOnInit() {
    let seekBar = <android.widget.SeekBar>this.el
                  .nativeElement.android;
    seekBar.setOnTouchListener(
      new android.view.View.OnTouchListener({
        onTouch(view, event) {
          return true;
        }
      })
    );
    seekBar.getThumb().mutate().setAlpha(0);

}
}
```

We gain access to the native `android.widget.SeekBar` instance via the `android` getter on the `Slider` component. We use Android's API reference documentation (`https://developer.android.com/reference/android/ widget/SeekBar.html`) to locate the SeekBar's API and disable user interaction by overriding `OnTouchListener`, and we hide the thumb by setting its Drawable alpha to 0.

Now, create `app/modules/player/directives/slider.directive.d.ts`:

```
export declare class SlimSliderDirective { }
```

This will allow us to import and use our `SlimSlider` class as a standard ES6 module; Create `app/modules/player/directives/index.ts`:

```
import { SlimSliderDirective } from './slider.directive';

export const DIRECTIVES: any[] = [

SlimSliderDirective
];
```

At runtime, NativeScript will only build the appropriate platform-specific files into the target platform, completely excluding nonapplicable code. This is a very powerful way to create platform-specific functionality in your codebase.

To finish up, let's just ensure our directives are declared in `PlayerModule` at `app/modules/player/player.module.ts` with the following changes:

```
...
import { DIRECTIVES } from './directives';
...

@NgModule({
  ...
  declarations: [
    ...COMPONENTS,
    ...DIRECTIVES
  ],
  ...
})
export class PlayerModule { }
```

We should now see this on iOS with our playback paused at 6 seconds:

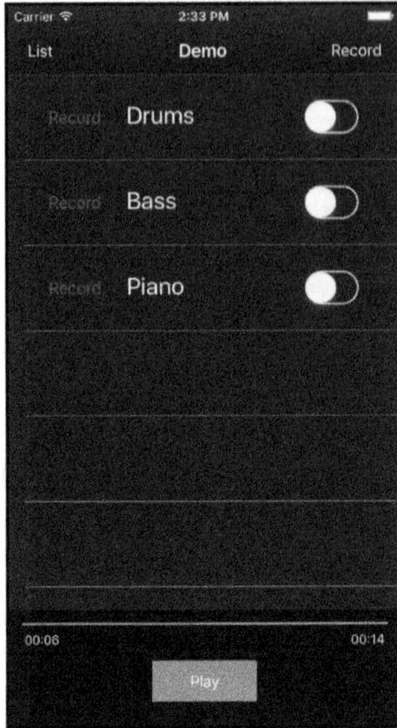

For Android, it will be as follows:

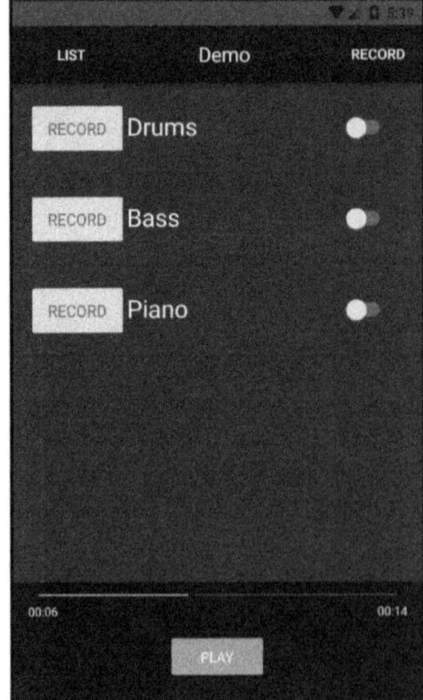

You can now observe the following:

- All the three tracks play together in a perfect mix
- Playback can be shuttled via the slider whether it's playing or not
- The play/pause toggle
- When playback reaches the end, our controls properly reset

And it all works on iOS and Android. An amazing feat, without question.

Summary

We are now fully immersed in the rich world of NativeScript, having introduced plugin integration as well as direct access to native APIs on iOS and Android. To top it off, we have a really neat multitrack player with full playback control, including shuttling through the mix!

The exciting combination of Angular, including its RxJS observable underpinnings, is really starting to shine through, where we've been able to take advantage of view bindings where needed and react to service event streams with powerful observable compositions, all while still retaining the ability to manually control our UI with fine grain control. Whether our view needs an Angular directive to enrich its capabilities or manual touch gesture control via raw NativeScript capabilities, we have it all at our fingertips now.

The fact that all along we are building a fully native iOS and Android app is truly mind blowing.

In the next chapter, we will continue to dig further into native APIs and plugins as we bring recording into our app's abilities to deliver on the core requirements of our multi-track recording studio mobile app.

8

Building an Audio Recorder

Recording audio is the most performance-intensive operation our app must handle. It is also the one feature where having access to native APIs will be the most rewarding. We want our users to be able to record with the lowest latency possible for the mobile device in order to achieve the highest fidelity of sound. Additionally, this recording should optionally happen over the top of an existing mix of pre-recorded tracks all playing in sync .

Since this phase of our app development will dive the deepest into platform-specific native APIs, we will split our implementations into two phases. We will first build out the iOS-specific details of the recording features, followed by Android.

In this chapter, we will cover the following:

- Building a feature rich cross-platform audio recorder for iOS and Android with a consistent API
- Integrating iOS framework libraries, such as AudioKit (`http://audiokit.io`), which was built entirely with Swift
- How to convert Swift/Objective C methods to NativeScript
- Building custom reusable NativeScript view components based on native APIs, as well as how to use them inside Angular
- Configuring a reusable Angular Component that can both be used via routing and opened via a popup modal
- Integrate Android Gradle libraries
- How to convert Java methods to NativeScript
- Using multiple item templates with NativeScript's ListView

Phase 1 – Building an audio recorder for iOS

The audio capabilities of the iOS platform are impressive, to say the least. A group of wonderfully talented audiophiles and software engineers have collaborated on building an open source framework layer on top of the platform's audio stack. This world class engineering effort is the awe inspiring AudioKit (`http://audiokit.io/`), led by the fearless Aurelius Prochazka, a true pioneer in audio technology.

The AudioKit framework is written entirely with Swift, which introduces a couple of interesting surface-level challenges when integrating with NativeScript.

Challenge detour – Integrate Swift based library into NativeScript

At the time of this writing, NativeScript can work with Swift if the codebase properly exposes the classes and types to Objective-C via what's called a **bridging header**, allowing both the languages to be mixed or matched. You can learn more about what a bridging header is here: `https://developer.apple.com/library/content/documentation/Swift/Conceptual/BuildingCocoaApps/MixandMatch.html`.
This bridging header is auto generated when the Swift codebase is compiled into a framework. Swift offers rich language features, some of which do not have a direct correlation to Objective C. Full featured support for the latest Swift language enhancements will likely come to NativeScript eventually however at the time of this writing there are a couple considerations to keep in mind.

AudioKit utilizes the best of what the Swift language has to offer, including enriched **enum** capabilities. You can learn more about the expanded enum features in the Swift language here:
`https://developer.apple.com/library/content/documentation/Swift/Conceptual/Swift_Programming_Language/Enumerations.html`

In particular, there is this from the documentation: "*they adopt many features traditionally supported only by classes, such as computed properties to provide additional information about the enumeration's current value, and instance methods to provide functionality related to the values the enumeration represents.*

Such *enums* are foreign to Objective C and, therefore, cannot be made available in the bridging header. Any code that uses Swift's exotic *enums* will be simply ignored when the bridging header is generated at compile time, resulting in Objective C not being able to interact with those sections of the code. This means you will not be able to use a method from a Swift codebase in NativeScript which utilizes these enhanced constructs out of the box (*at the time of this writing*).

To remedy this, we will fork the AudioKit framework and flatten the exotic enums used in the `AKAudioFile` extension files, which provide a powerful and convenient export method we will want to use to save our recorded audio files. The exotic *enum* we need to modify looks like this (`https://github.com/audiokit/AudioKit/blob/master/AudioKit/Common/Internals/Audio%20File/AKAudioFile%2BProcessingAsynchronously.swift`):

```
// From AudioKit's Swift 3.x codebase

public enum ExportFormat {
  case wav
  case aif
  case mp4
  case m4a
  case caf

  fileprivate var UTI: CFString {
    switch self {
    case .wav:
      return AVFileTypeWAVE as CFString
    case .aif:
      return AVFileTypeAIFF as CFString
    case .mp4:
      return AVFileTypeAppleM4A as CFString
    case .m4a:
      return AVFileTypeAppleM4A as CFString
    case .caf:
      return AVFileTypeCoreAudioFormat as CFString
    }
  }

  static var supportedFileExtensions: [String] {
    return ["wav", "aif", "mp4", "m4a", "caf"]
  }
}
```

This is unlike any *enum* you may be familiar with; as you can see, it includes properties in addition to what enums have. When this code is compiled and the bridging header is generated to mix or match with Objective-C, the bridging header will then exclude any code that uses this construct. We will flatten this out to look like the following:

```
public enum ExportFormat: Int {
  case wav
  case aif
  case mp4
  case m4a
  case caf
}

static public func stringUTI(type: ExportFormat) -> CFString {
  switch type {
  case .wav:
    return AVFileTypeWAVE as CFString
  case .aif:
    return AVFileTypeAIFF as CFString
  case .mp4:
    return AVFileTypeAppleM4A as CFString
  case .m4a:
    return AVFileTypeAppleM4A as CFString
  case .caf:
    return AVFileTypeCoreAudioFormat as CFString
  }
}

static public var supportedFileExtensions: [String] {
  return ["wav", "aif", "mp4", "m4a", "caf"]
}
```

We will then adjust the portions of the `AKAudioFile` extension to use our flattened properties. This will allow us to manually build `AudioKit.framework` we can use in our app, exposing the method we want to use: `exportAsynchronously`.

We won't go over the details of manually building `AudioKit.framework`, as it is well documented here: `https://github.com/audiokit/AudioKit/blob/master/Frameworks/INSTALL.md#building-universal-frameworks-from-scratch`. With our custom-built framework, we are now ready to integrate it into our app.

Integrating a custom-built iOS framework into NativeScript

We can now create an internal plugin to integrate this iOS framework into our app. Take the custom `AudioKit.framework` we have built and create a `nativescript-audiokit` directory at the root of our app. We then add a `platforms/ios` folder inside to drop the framework into. This will let NativeScript know how to build these iOS-specific files into the app. As we want this internal plugin to be treated like any standard npm plugin, we will also add `package.json` directly inside the `nativescript-audiokit` folder with the following contents:

```
{
  "name": "nativescript-audiokit",
  "version": "1.0.0",
  "nativescript": {
    "platforms": {
      "ios": "3.0.0"
    }
  }
}
```

We will now use the following command to add it to our app (NativeScript will look locally first and find the **nativescript-audiokit** plugin):

```
tns plugin add nativescript-audiokit
```

This will properly add the custom-built iOS framework into our app.
However, we need two more very important items:

1. Since AudioKit is a Swift-based framework, we want to ensure our app includes the proper supporting Swift libraries. Add a new file, `nativescript-audiokit/platforms/ios/build.xcconfig`:

   ```
   EMBEDDED_CONTENT_CONTAINS_SWIFT = true
   ```

2. Since we will be engaging with the user's microphone, we will want to ensure the microphone usage is indicated in our app's property list. We will also take this opportunity to add two additional property settings to enhance our app's abilities. So, in total, we will add three property keys for the following purposes:
   - Let the device know our app needs access to the microphone and ensure the user's permission is requested on first access.
   - Continue playing audio if the app is placed into the background.

- Provide the ability to see the app's `documents` folder in iTunes when the phone is connected to a computer. This will allow you to view recorded files right inside of iTunes via the app's Documents. This could be useful for integration into a desktop audio editing software.

Add a new file, `nativescript-audiokit/platforms/ios/Info.plist`, with the following code:

```
<?xml version="1.0" encoding="UTF-8"?>
<!DOCTYPE plist PUBLIC "-//Apple//DTD PLIST 1.0//EN"
"http://www.apple.com/DTDs/PropertyList-1.0.dtd">
<plist version="1.0">
<dict>
  <key>NSMicrophoneUsageDescription</key>
  <string>Requires access to microphone.</string>
  <key>UIBackgroundModes</key>
  <array>
    <string>audio</string>
  </array>
  <key>UIFileSharingEnabled</key>
  <true/>
</dict>
</plist>
```

Here is a screenshot to better illustrate the internal plugin structure in our app:

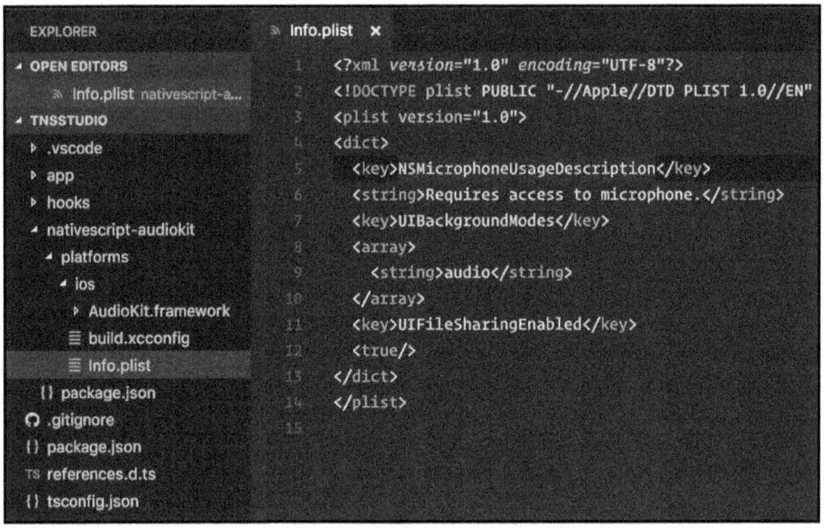

Now, when NativeScript builds the iOS app, it will ensure `AudioKit.framework` is included as a library and merge the contents of `build.xcconfig` and `Info.plist` into our app's configuration. Any time we make changes to the files inside this internal plugin folder (`nativescript-audiokit`), we want to ensure our app picks up those changes. To do so, we can simply remove and add the plugin back, so let's do that now:

```
tns plugin remove nativescript-audiokit
tns plugin add nativescript-audiokit
```

We are now ready to build our audio recorder using the AudioKit API for iOS.

Setting up native API type checking and generate AudioKit TypeScript definitions

The first thing we want to do is install `tns-platform-declarations`:

```
npm i tns-platform-declarations --save-dev
```

Now, we create a new file in the root of the project called `references.d.ts` with the following contents:

```
/// <reference path="./node_modules/tns-platform-declarations/ios.d.ts" />
/// <reference path="./node_modules/tns-platform-declarations/android.d.ts"
/>
```

This provides us with full type checking and intellisense support for iOS and Android APIs.

We now want to generate typings for the AudioKit framework itself. We can execute this command to generate the typings for the included `AudioKit.framework`:

```
TNS_TYPESCRIPT_DECLARATIONS_PATH="$(pwd)/typings" tns build ios
```

We are setting the environment variable `TNS_TYPESCRIPT_DECLARATIONS_PATH` to the present working directory (`pwd`) with a folder prefix of `typings`. When NativeScript creates the iOS build, it will also generate type definition files for all the native APIs available to our app, including third-party libraries. We will now see a `typings` folder appear in our project, containing two folders: `i386` and `x86_64`. One is for the Simulator architecture and the other the device. Both will contain the same output, so we can just focus on one. Open the `i386` folder and you will find an `objc!AudioKit.d.ts` file.

We want to use only that file, so move it to the root of the `typings` folder: `typings/objc!AudioKit.d.ts`. We can then remove both the `i386` and `x86_64` folders, as we will no longer need them (the other API definition files are provided via `tns-platform-declarations`). We just generated these typings to get TypeScript definitions for the AudioKit library. This is a one-time thing, done to integrate easily with this native library, so you are safe to add this custom `typings` folder to source control.

Double-check `tsconfig.json` and ensure you have the `"skipLibCheck": true` option enabled. We can now modify our `references.d.ts` file to include the additional types for the AudioKit library:

```
/// <reference path="./node_modules/tns-platform-declarations/ios.d.ts" />
/// <reference path="./node_modules/tns-platform-declarations/android.d.ts"
/>
/// <reference path="./typings/objc!AudioKit.d.ts" />
```

Our project structure should now look like this:

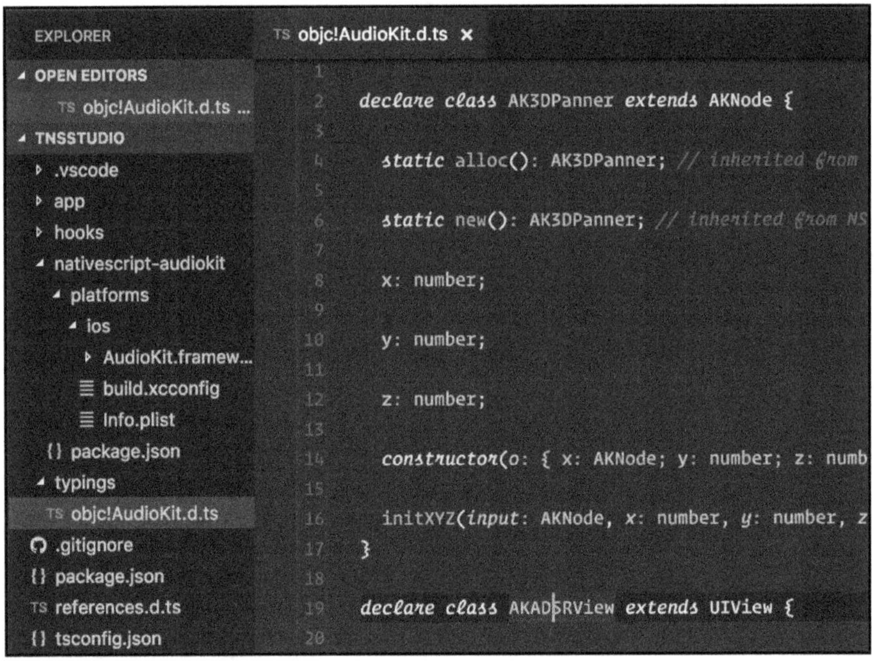

Build recorder with AudioKit

We will begin by creating a model around our interaction with AudioKit's recording APIs. You could just start writing directly against these APIs right from your Angular component or service, but since we want to provide a consistent API across iOS and Android, there's a smarter way to architect this. Instead, we will abstract a simple API, usable across both platforms, which will tap into the correct native implementations under the hood.

There will be a lot of interesting details related to AudioKit going on here, but create `app/modules/recorder/models/record.model.ts` with the following and we will explain some of the bits in a moment:

Later, we will add the `.ios.ts` suffix to this model, since it will contain iOS-specific implementation details. However, here in Phase 1, we will use the model directly (omitting the platform suffix) while we develop our iOS recorder.

```
import { Observable } from 'data/observable';
import { knownFolders } from 'file-system';

// all available states for the recorder
export enum RecordState {
  readyToRecord,
  recording,
  readyToPlay,
  playing,
  saved,
  finish
}

// available events
export interface IRecordEvents {
  stateChange: string;
}

// for use when saving files
const documentsFilePath = function(filename: string) {
  return `${knownFolders.documents().path}/${filename}`;
}

export class RecordModel extends Observable {

  // available events to listen to
  private _events: IRecordEvents;

  // control nodes
```

```
private _mic: AKMicrophone;
private _micBooster: AKBooster;
private _recorder: AKNodeRecorder;

// mixers
private _micMixer: AKMixer;
private _mainMixer: AKMixer;

// state
private _state: number = RecordState.readyToRecord;

// the final saved path to use
private _savedFilePath: string;

constructor() {
  super();
  // setup the event names
  this._setupEvents();

  // setup recording environment
  // clean any tmp files from previous recording sessions
  (<any>AVAudioFile).cleanTempDirectory();

  // audio setup
  AKSettings.setBufferLength(BufferLength.Medium);

  try {
    // ensure audio session is PlayAndRecord
    // allows mixing with other tracks while recording
    AKSettings.setSessionWithCategoryOptionsError(
      SessionCategory.PlayAndRecord,
      AVAudioSessionCategoryOptions.DefaultToSpeaker
    );
  } catch (err) {
    console.log('AKSettings error:', err);
  }

  // setup mic with it's own mixer
  this._mic = AKMicrophone.alloc().init();
  this._micMixer = AKMixer.alloc().init(null);
  this._micMixer.connect(this._mic);
  // Helps provide mic monitoring when headphones are plugged in
  this._micBooster = AKBooster.alloc().initGain(<any>this._micMixer, 0);

  try {
    // recorder takes the micMixer input node
    this._recorder = AKNodeRecorder.alloc()
      .initWithNodeFileError(<any>this._micMixer, null);
```

```
    } catch (err) {
      console.log('AKNodeRecorder init error:', err);
    }

    // overall main mixer uses micBooster
    this._mainMixer = AKMixer.alloc().init(null);
    this._mainMixer.connect(this._micBooster);

    // single output set to mainMixer
    AudioKit.setOutput(<any>this._mainMixer);
    // start the engine!
    AudioKit.start();
  }

  public get events(): IRecordEvents {
    return this._events;
  }

  public get mic(): AKMicrophone {
    return this._mic;
  }

  public get recorder(): AKNodeRecorder {
    return this._recorder;
  }

  public get audioFilePath(): string {
    if (this._recorder) {
      return this._recorder.audioFile.url.absoluteString;
    }
    return '';
  }

  public get state(): number {
    return this._state;
  }

  public set state(value: number) {
    this._state = value;
    // always emit state changes
    this._emitEvent(this._events.stateChange, this._state);
  }

  public get savedFilePath() {
    return this._savedFilePath;
  }

  public set savedFilePath(value: string) {
```

```
    this._savedFilePath = value;
    if (this._savedFilePath)
      this.state = RecordState.saved;
  }

  public toggleRecord() {
    if (this._state !== RecordState.recording) {
      // just force ready to record
      // when coming from any state other than recording
      this.state = RecordState.readyToRecord;

      if (this._recorder) {
        try {
          // resetting (clear previous recordings)
          this._recorder.resetAndReturnError();
        } catch (err) {
          console.log('Recorder reset error:', err);
        }
      }
    }

    switch (this._state) {
      case RecordState.readyToRecord:
        if (AKSettings.headPhonesPlugged) {
          // Microphone monitoring when headphones plugged
          this._micBooster.gain = 1;
        }

        try {
          this._recorder.recordAndReturnError();
          this.state = RecordState.recording;
        } catch (err) {
          console.log('Recording failed:', err);
        }
        break;
      case RecordState.recording:
        this.state = RecordState.readyToPlay;
        this._recorder.stop();
        // Microphone monitoring muted when playing back
        this._micBooster.gain = 0;
        break;
    }
  }

  public togglePlay() {
    if (this._state === RecordState.readyToPlay) {
      this.state = RecordState.playing;
    } else {
```

```
        this.stopPlayback();
      }
    }

    public stopPlayback() {
      if (this.state !== RecordState.recording) {
        this.state = RecordState.readyToPlay;
      }
    }

    public save() {
      let fileName = `recording-${Date.now()}.m4a`;
      this._recorder.audioFile
  .exportAsynchronouslyWithNameBaseDirExportFormatFromSampleToSampleCallback(
        fileName, BaseDirectory.Documents, ExportFormat.M4a, null, null,
        (af: AKAudioFile, err: NSError) => {
          this.savedFilePath = documentsFilePath(fileName);
        });
    }

    public finish() {
      this.state = RecordState.finish;
    }

    private _emitEvent(eventName: string, data?: any) {
      let event = {
        eventName,
        data,
        object: this
      };
      this.notify(event);
    }

    private _setupEvents() {
      this._events = {
        stateChange: 'stateChange'
      };
    }
  }
```

`RecordModel` will behave a bit like a state machine, where the only states it could be in are the following:

- `readyToRecord`: Default starting state. Must be in this state to enter the recording state.
- `recording`: Quiet in the studio! Recording in process.

- `readyToPlay`: User has stopped recording and now has a recorded file to play back with the mix.
- `playing`: User is playing back the recorded file with the mix.
- `saved`: User chose to save the recording, that should kick off actions to save the new track with the active composition.
- `finish`: Once the save actions are complete, the recorder should shut down.

We then define the shape of the events the recorder will provide with `IRecordEvents`. In this case, we will have a single event, `stateChange`, which will notify any listeners when the state changes (*see the state setter*). Our model will extend NativeScript's `Observable` class (hence, `RecordModel extends Observable`), which will provide us with the notify API to dispatch our events.

We then set up several references to the various AudioKit bits we will use. Most of what is designed is directly from this AudioKit example on recording: `https://github.com/audiokit/AudioKit/blob/master/Examples/iOS/RecorderDemo/RecorderDemo/ViewController.swift`. We even use the same state enum setup (with a few extras). In their example, AudioKit's `AKAudioPlayer` is used for playback; but, with our design, we will load our recorded files into our multitrack player design to play them back with our mix. We could work `AKAudioPlayer` into `TrackPlayerModel` for iOS; but, `TNSPlayer` (from the **nativescript-audio** plugin) is cross-platform compatible and will work just fine. We'll cover the details of how we load these new recorded files into our design shortly, but notifying listeners of the recorder's state will provide us all the flexibility we need to handle all that when we get there.

You may wonder why we type-cast this:

```
(<any>AVAudioFile).cleanTempDirectory();
```

Good question. AudioKit provides Extensions to Core Foundation classes such as `AVAudioFile`. These were known as `Categories` in Objective C: `https://developer.apple.com/library/content/documentation/General/Conceptual/DevPedia-CocoaCore/Category.html`; however, in Swift, they are referred to as `Extensions`: `https://developer.apple.com/library/content/documentation/Swift/Conceptual/Swift_Programming_Language/Extensions.html`.

If you recall, we generated TypeScript definitions for AudioKit; but, we only kept the `objc!AudioKit.d.ts` file to reference. If we had looked in the foundation definitions, we would have seen the extension to `AVAudioFile`. However, since we did not keep those definitions around and instead are relying on the default `tns-platform-declarations` definitions, this `Extension` is not known to our TypeScript compiler, so we simply type-cast it, as we know AudioKit provides this.

It's also critical `RecordModel` sets the audio session to `PlayAndRecord`, as this will allow us to record while playing our mix at the same time:

```
AKSettings.setSessionWithCategoryOptionsError(
  SessionCategory.PlayAndRecord,
  AVAudioSessionCategoryOptions.DefaultToSpeaker
);
```

You may also be curious why some classes use `init()` and others `init(null)`:

```
this._mic = AKMicrophone.alloc().init();
this._micMixer = AKMixer.alloc().init(null);
this._micMixer.connect(this._mic);
```

Some of the initializers for AudioKit's classes take an optional argument, for example, `AKMixer` takes an optional `NSArray` of `AVAudioNode` to connect. However, our TypeScript definitions have those defined as required, so we are just passing `null` to that argument and instead using the `connect` node API directly.

How to convert Swift/ObjC methods to NativeScript

The last point of interest from `RecordModel` might be the `save` method, which will export our recording from the app's `tmp` directory to the app's `documents` folder while converting it to the smaller filesize `.m4a` audio format:

```
this._recorder.audioFile
.exportAsynchronouslyWithNameBaseDirExportFormatFromSampleToSampleCallback(
    fileName, BaseDirectory.Documents, ExportFormat.M4a, null, null,
    (af: AKAudioFile, err: NSError) => {
      this.savedFilePath = documentsFilePath(fileName);
  });
```

Long method name, right? Yes, indeed; some Swift/ObjC parameterized method names collapse to become very long. That particular method in Swift is defined as follows:

```
exportAsynchronously(name:baseDir:exportFormat:fromSample:toSample:callback
:)
// converted to NativeScript:
exportAsynchronouslyWithNameBaseDirExportFormatFromSampleToSampleCallback
```

Since we had the TypeScript definitions generated for AudioKit, they helped us out here. However, sometimes, you don't have that luxury. A Swift/ObjC method with various parameters for arguments collapse into each other while adding `With` in between the start of the method name and the start of the parameter argument names, while capitalizing the first character upon collapsing each.

Building custom reusable NativeScript view for native audio Waveform display

Instead of creating an Angular Component for our Waveform display, we will create a custom NativeScript view component, that taps into native APIs, that we can then register with Angular to use in our components. The reason for doing this is due to NativeScript's powerful `view` base class that we can extend, which provides a nice API when using underlying native APIs for the `view`. This Waveform display will work in tandem with the `RecordModel` we just created to bring to life our real-time Waveform feedback display of the device's microphone. It would also be amazing to reuse this Waveform display as a static audio file waveform rendering on our track list, as an alternate view for our main composition view. AudioKit provides classes and APIs to do all this.

Since we want to be able to use this anywhere in our app, we will create it inside the shared module directory; however, keep in mind that it could live anywhere. It doesn't matter so much here, since this is not an Angular component that needs to be declared in `NgModule`. Additionally, since this will specifically work with native APIs, let's create it inside a new `native` folder to potentially house other NativeScript-specific view components.

Create `app/modules/shared/native/waveform.ts` with the following contents, which we will explain in a moment:

```
import { View, Property } from 'ui/core/view';
import { Color } from 'color';

// Support live microphone display as well as static audio file renders
type WaveformType = 'mic' | 'file';

// define properties
export const plotColorProperty = new Property<Waveform, string>({ name:
'plotColor' });
export const plotTypeProperty = new Property<Waveform, string>({ name:
'plotType' });
export const fillProperty = new Property<Waveform, string>({ name: 'fill'
});
export const mirrorProperty = new Property<Waveform, string>({ name:
'mirror' });

export interface IWaveformModel {
  readonly target: any;
  dispose(): void;
}
export class Waveform extends View {
  private _model: IWaveformModel;
```

```
  private _type: WaveformType;

  public set type(value: WaveformType) {
    this._type = value;
  }

  public get type() {
    return this._type;
  }

  public set model(value: IWaveformModel) {
    this._model = value;
  }

  public get model() {
    return this._model;
  }

  createNativeView() {
    switch (this.type) {
      case 'mic':
        this.nativeView = AKNodeOutputPlot.alloc()
          .initFrameBufferSize(this._model.target, CGRectMake(0, 0, 0, 0),
1024);
        break;
      case 'file':
        this.nativeView = EZAudioPlot.alloc().init();
        break;
    }
    return this.nativeView;
  }

  initNativeView() {
    if (this._type === 'file') {
      // init file with the model's target
      // target should be absolute url to path of file
      let file = EZAudioFile.alloc()
        .initWithURL(NSURL.fileURLWithPath(this._model.target));
      // render the file's data as a waveform
      let data = file.getWaveformData();
      (<EZAudioPlot>this.nativeView)
        .updateBufferWithBufferSize(data.buffers[0], data.bufferSize);
    }
  }

  disposeNativeView() {
    if (this.model && this.model.dispose) this.model.dispose();
  }
```

```
    [plotColorProperty.setNative](value: string) {
      this.nativeView.color = new Color(value).ios;
    }

    [fillProperty.setNative](value: string) {
      this.nativeView.shouldFill = value === 'true';
    }

    [mirrorProperty.setNative](value: string) {
      this.nativeView.shouldMirror = value === 'true';
    }

    [plotTypeProperty.setNative](value: string) {
      switch (value) {
        case 'buffer':
          this.nativeView.plotType = EZPlotType.Buffer;
          break;
        case 'rolling':
          this.nativeView.plotType = EZPlotType.Rolling;
          break;
      }
    }
  }

  // register properties with it's type
  plotColorProperty.register(Waveform);
  plotTypeProperty.register(Waveform);
  fillProperty.register(Waveform);
  mirrorProperty.register(Waveform);
```

We are creating several properties using NativeScript's `Property` class, which will add great conveniences when exposing native view properties through the view binding properties. One such convenience in defining these properties with the `Property` class, these setters will only be called when `nativeView` is defined, avoiding double invoked property setters (one via a pure JS property setter, which is the alternative, and potentially another for when the underlying `nativeView` is ready).

When wanting to expose native view properties that could be bound via your custom component, define several `Property` classes for them, referencing the name you'd like to use for the view binding.

```
  // define properties
  export const plotColorProperty = new Property<Waveform, string>({ name:
  'plotColor' });
  export const plotTypeProperty = new Property<Waveform, string>({ name:
  'plotType' });
```

```
export const fillProperty = new Property<Waveform, string>({ name: 'fill'
});
export const mirrorProperty = new Property<Waveform, string>({ name:
'mirror' });
```

By setting up these `Property` instances, we can now do this in our view component class:

```
[plotColorProperty.setNative](value: string) {
  this.nativeView.color = new Color(value).ios;
}
```

This will only be invoked once `nativeView` is ready, which is exactly what we want. You can read more about this particular syntax and notation in this draft written by core team member Alex Vakrilov: `https://gist.github.com/vakrilov/ca888a1ea410f4ea7a4c7b2035e06b07#registering-the-property`.

Then, at the bottom of our class (after it's defined), we register the class with the `Property` instances:

```
// register properties
plotColorProperty.register(Waveform);
plotTypeProperty.register(Waveform);
fillProperty.register(Waveform);
mirrorProperty.register(Waveform);
```

Okay, with that explained, let's look at some other elements to this implementation.

We are also introducing a helpful interface here, which we will apply to `RecordModel` in a moment:

```
export interface IWaveformModel {
  readonly target: any;
  dispose(): void;
}
```

This will help define a shape for other models to implement, ensuring they conform to an API the Waveform display expects:

- `target`: Defines the key input to be used with the native class.
- `dispose()`: Each model should provide this method to handle any clean up when the view is destroyed.

This is the custom NativeScript 3.x View Life cycle call execution order:

1. `createNativeView():AnyNativeView;` // Create your native view.
2. `initNativeView():void;` // Init your native view.
3. `disposeNativeView():void;` // Clean up your native view.

The `createNativeView` method overridden from NativeScript's `View` class is likely the most interesting:

```
createNativeView() {
  switch (this.type) {
    case 'mic':
      this.nativeView = AKNodeOutputPlot.alloc()
        .initFrameBufferSize(this._model.target, CGRectMake(0, 0, 0, 0),
1024);
      break;
    case 'file':
      this.nativeView = EZAudioPlot.alloc().init();
      break;
  }
  return this.nativeView;
}
```

Here, we allow the `type` property to determine which type of Waveform display it should render.
In the case of `mic`, we utilize AudioKit's `AKNodeOutputPlot` (which actually extends `EZAudioPlot` under the hood) to initialize a waveform (that is, `audioplot`) using our model's target, which will end up being our RecordModel's microphone.
In the case of `file`, we utilize AudioKit's `EZAudioPlot` directly to create a static waveform representing an audio file.

The `initNativeView` method, also overridden from NativeScript's `View` class, is called second in its life cycle and provides a way to initialize your native view. You might find it interesting that we call the setters again here. The setters are called first when the component bindings are set via the XML and the class instantiates, which is *before* `createNativeView` and `initNativeView` are called. This why we cache the values in private references. However, we also want these setters to modify `nativeView` with Angular's view bindings (when changed dynamically), which is why we also have `if (this.nativeView)` inside the setters to change `nativeView` dynamically when available.

The `disposeNativeView` method (you guessed it, also overridden from the `{N}` of the `View` class) is called when `View` gets destroyed, which is where we call the model's `dispose` method if available.

Integrate a custom NativeScript view into our Angular app

To use our NativeScript Waveform view within Angular, we need to first register it. You can do this in the root module, root app component, or another place that is initialized at boot time (usually, not in a lazy-loaded module). To be tidy, we will register it within `SharedModule` in the same directory, so add the following in `app/modules/shared/shared.module.ts`:

```
...
// register nativescript custom components
import { registerElement } from 'nativescript-angular/element-registry';
import { Waveform } from './native/waveform';
registerElement('Waveform', () => Waveform);
...
@NgModule({...
export class SharedModule {...
```

The `registerElement` method allows us to define the name of the Component we want to use within Angular components as the first argument, and takes a resolver function that should return the NativeScript `View` class to use for it.

Let's now use our new `IWaveformModel` and clean up some of `RecordModel` to use it, as well as prepare to create our Android implementation next. Let's refactor a couple things out of `RecordModel` into a common file to share code between our iOS and Android (coming soon!) models.

Create `app/modules/recorder/models/record-common.ts`:

```
import { IWaveformModel } from '../../shared/native/waveform';
import { knownFolders } from 'file-system';

export enum RecordState {
  readyToRecord,
  recording,
  readyToPlay,
  playing,
  saved,
  finish
```

```
}

export interface IRecordEvents {
  stateChange: string;
}

export interface IRecordModel extends IWaveformModel {
  readonly events: IRecordEvents;
  readonly recorder: any;
  readonly audioFilePath: string;
  state: number;
  savedFilePath: string;
  toggleRecord(): void;
  togglePlay(startTime?: number, when?: number): void;
  stopPlayback(): void;
  save(): void;
  finish(): void;
}

export const documentsFilePath = function(filename: string) {
  return `${knownFolders.documents().path}/${filename}`;
}
```

This contains most of what was at the top of `RecordModel`, with the addition of the `IRecordModel` interface, which extends `IWaveformModel`. Since we built out our iOS implementation, we now have a model shape we would like our Android implementation to adhere to. Abstracting that shape into an interface will provide us a clear path to follow when we move to Android momentarily.

For convenience, let's also create an index for our models, which would also expose this common file, in `app/modules/recorder/models/index.ts`:

```
export * from './record-common.model';
export * from './record.model';
```

We can now modify `RecordModel` to import these common items, as well as implement this new `IRecordModel` interface. Since this new interface also *extends* `IWaveformModel`, it will immediately tell us we need to implement the `readonly target` getter and the `dispose()` method, as required to be used with our Waveform view:

```
import { Observable } from 'data/observable';
import { IRecordModel, IRecordEvents, RecordState, documentsFilePath } from
'./common';

export class RecordModel extends Observable implements IRecordModel {
  ...
  public get target() {
```

```
    return this._mic;
  }

  public dispose() {
    AudioKit.stop();
    // cleanup
    this._mainMixer = null;
    this._recorder = null;
    this._micBooster = null;
    this._micMixer = null;
    this._mic = null;
    // clean out tmp files
    (<any>AVAudioFile).cleanTempDirectory();
  }
  ...
```

The `target` of `RecordModel` will be the microphone that the Waveform view will use. Our `dispose` method will stop the AudioKit engine while doing reference clean up, as well as ensuring to clean out any temporary files created while recording.

Creating the Recorder View layout

When the user taps on **Record** in the top right corner of the app, it prompts the user to authenticate, after which the app routes to the record view. Additionally, it would be nice to reuse this record view in a modal popup to show when the composition contains tracks, so the user doesn't feel like they are leaving the composition while recording. However, when the composition is new, it's fine to navigate to the record view via routing. We will show how this can be done, but let's first set up our layout using the new fancy Waveform view and our powerful new `RecordModel`.

Add the following to `app/modules/recorder/components/record.component.html`:

```
<ActionBar title="Record" icon="" class="action-bar">
  <NavigationButton visibility="collapsed"></NavigationButton>
  <ActionItem text="Cancel"
    ios.systemIcon="1" android.systemIcon="ic_menu_back"
    (tap)="cancel()"></ActionItem>
</ActionBar>
<FlexboxLayout class="record">
  <GridLayout rows="auto" columns="auto,*,auto" class="p-10"
*ngIf="isModal">
    <Button text="Cancel" (tap)="cancel()"
      row="0" col="0" class="c-white"></Button>
  </GridLayout>
  <Waveform class="waveform"
```

```
      [model]="recorderService.model"
      type="mic"
      plotColor="yellow"
      fill="false"
      mirror="true"
      plotType="buffer">
    </Waveform>
    <StackLayout class="p-5">
      <FlexboxLayout class="controls">
        <Button text="Rewind" class="btn text-center"
          (tap)="recorderService.rewind()"
          [isEnabled]="state == recordState.readyToPlay || state ==
recordState.playing">
        </Button>
        <Button [text]="recordBtn" class="btn text-center"
          (tap)="recorderService.toggleRecord()"
          [isEnabled]="state != recordState.playing"></Button>
        <Button [text]="playBtn" class="btn text-center"
          (tap)="recorderService.togglePlay()"
          [isEnabled]="state == recordState.readyToPlay || state ==
recordState.playing">
        </Button>
      </FlexboxLayout>
      <FlexboxLayout class="controls bottom"
        [class.recording]="state == recordState.recording">
        <Button text="Save" class="btn"
          [class.save-ready]="state == recordState.readyToPlay"
          [isEnabled]="state == recordState.readyToPlay"
          (tap)="recorderService.save()"></Button>
      </FlexboxLayout>
    </StackLayout>
  </FlexboxLayout>
```

We are using `FlexboxLayout` because we want our Waveform view to stretch to cover the full available vertical space, leaving only the recorder's controls positioned at the bottom. `FlexboxLayout` is a very versatile layout container, which provides most of the same CSS styling attributes found with the the flexbox model on the web.

Interestingly, we show a **Cancel** button inside a `GridLayout` container only when displayed as a modal, since we need a way to close the modal. ActionBars are ignored and not displayed when the view is opened via a modal.

ActionBars are ignored when the view is opened via a modal, so they are not displayed in the modal. `ActionBar` is shown on navigated views only.

Furthermore, our `ActionBar` setup is rather interesting here and is one of the areas of NativeScript view layouts where iOS and Android differ the most. On iOS, `NavigationButton` has a default behavior, that automatically pops the view from the stack and animates back to the previous view. Additionally, any tap events on `NavigationButton` on iOS are completely ignored, whereas on Android, the tap event is triggered on `NavigationButton`. Because of this crucial difference, we want to completely ignore `NavigationButton` of `ActionBar` by using `visibility="collapsed"` to ensure it is never shown. Instead, we use `ActionItem` with an explicit tap event to ensure the correct logic is triggered on our component for both platforms.

`NavigationButton` behavior on iOS and Android is different:

- **iOS**: `NavigationButton` ignores (tap) events, and this button appears by default when navigating to a view.
- **Android**: `NavigationButton` (tap) events are triggered.

You can see our Waveform (the custom NativeScript) view in use here. We use Angular's binding syntax when binding the model, since it's an object. For the other properties, we specify their values directly, since they are primitive values. We could, however, use Angular's binding syntax on those as well if we wanted to change those values dynamically via user interaction. For example, we could show a fun color picker, which would allow the user to change the color (`plotColor`) of the waveform on the fly.

We'll provide a component-specific stylesheet for our record component, `app/modules/recorder/components/record.component.css`:

```
.record {
  background-color: rgba(0,0,0,.5);
  flex-direction: column;
  justify-content: space-around;
  align-items: stretch;
  align-content: center;
}

.record .waveform {
  background-color: transparent;
  order: 1;
  flex-grow: 1;
}

.controls {
  width: 100%;
  height: 200;
```

```
    flex-direction: row;
    flex-wrap: nowrap;
    justify-content: center;
    align-items: center;
    align-content: center;
  }

  .controls.bottom {
    height: 90;
    justify-content: flex-end;
  }

  .controls.bottom.recording {
    background-color: #B0342D;
  }

  .controls.bottom .btn {
    border-radius: 40;
    height: 62;
    padding: 2;
  }

  .controls.bottom .btn.save-ready {
    background-color: #42B03D;
  }

  .controls .btn {
    color: #fff;
  }

  .controls .btn[isEnabled=false] {
    background-color: transparent;
    color: #777;
  }
```

Some of these CSS properties may look familiar if you've used the flexbox model on the web. A great and fun resource to learn more about flexbox styling is Flexbox Zombies by Dave Geddes: `http://flexboxzombies.com`.

At this point, our CSS is starting to grow and we could clean things up a lot with SASS. We will do exactly that, coming up soon, so hang in there!

Now, let's take a look at the Component at `app/modules/recorder/components/record.component.ts`:

```
// angular
import { Component, OnInit, OnDestroy, Optional } from '@angular/core';

// libs
import { Subscription } from 'rxjs/Subscription';

// nativescript
import { RouterExtensions } from 'nativescript-angular/router';
import { ModalDialogParams } from 'nativescript-
angular/directives/dialogs';
import { isIOS } from 'platform';

// app
import { RecordModel, RecordState } from '../models';
import { RecorderService } from '../services/recorder.service';

@Component({
  moduleId: module.id,
  selector: 'record',
  templateUrl: 'record.component.html',
  styleUrls: ['record.component.css']
})
export class RecordComponent implements OnInit, OnDestroy {
  public isModal: boolean;
  public recordBtn: string = 'Record';
  public playBtn: string = 'Play';
  public state: number;
  public recordState: any = {};

  private _sub: Subscription;

  constructor(
    private router: RouterExtensions,
    @Optional() private params: ModalDialogParams,
    public recorderService: RecorderService
  ) {
    // prepare service for brand new recording
    recorderService.setupNewRecording();

    // use RecordState enum names as reference in view
    for (let val in RecordState ) {
      if (isNaN(parseInt(val))) {
        this.recordState[val] = RecordState[val];
      }
    }
```

```
  }

  ngOnInit() {
    if (this.params && this.params.context.isModal) {
      this.isModal = true;
    }
    this._sub = this.recorderService.state$.subscribe((state: number) => {
      this.state = state;
      switch (state) {
        case RecordState.readyToRecord:
        case RecordState.readyToPlay:
          this._resetState();
          break;
        case RecordState.playing:
          this.playBtn = 'Pause';
          break;
        case RecordState.recording:
          this.recordBtn = 'Stop';
          break;
        case RecordState.finish:
          this._cleanup();
          break;
      }
    });
  }

  ngOnDestroy() {
    if (this._sub) this._sub.unsubscribe();
  }

  public cancel() {
    this._cleanup();
  }

  private _cleanup() {
    this.recorderService.cleanup();
    invokeOnRunLoop(() => {
      if (this.isModal) {
        this._close();
      } else {
        this._back();
      }
    });
  }

  private _close() {
    this.params.closeCallback();
  }
```

```
  private _back() {
    this.router.back();
  }

  private _resetState() {
    this.recordBtn = 'Record';
    this.playBtn = 'Play';
  }
}

/**
 * Needed on iOS to prevent this potential exception:
 * "This application is modifying the autolayout engine from a background
thread after the engine was accessed from the main thread. This can lead to
engine corruption and weird crashes."
 */
const invokeOnRunLoop = (function () {
  if (isIOS) {
    var runloop = CFRunLoopGetMain();
    return function(func) {
      CFRunLoopPerformBlock(runloop, kCFRunLoopDefaultMode, func);
      CFRunLoopWakeUp(runloop);
    }
  } else {
    return function (func) {
      func();
    }
  }
}());
```

Starting from the bottom of that file, you'll probably wonder what the heck `invokeOnRunLoop` is. This is a handy way to ensure thread safety in conditions where the thread might rear its ugly head. In this case, AudioKit's engine is started from the UI thread in `RecordModel`, since NativeScript marshals native calls on the UI thread. However, when our record view closes (whether it be from a modal our navigating back), some background threads are invoked. Wrapping our handling of closing this view with `invokeOnRunLoop` helps solve this transient exception. It's the answer to how to use iOS `dispatch_async(dispatch_get_main_queue(...))` with NativeScript.

Working our way up the file, we'll encounter `this.recorderService.state$.subscribe((state: number) =>` .... In a moment, we'll be implementing a way to observe the recording `state$` as an observable, so our view can simply react to its state changes.

Also of note, it is a handy way to collapse `RecordState enum` into properties we can use as view bindings to compare against the current state (`this.state = state;`).

When the component is constructed, `recorderService.setupNewRecording()` will prepare our service for a brand new recording each time this view appears.

Lastly, take note of the injection of `@Optional() private params: ModalDialogParams`. Earlier, we mentioned that *it would be nice to reuse this record view in a modal popup*. The interesting part is that `ModalDialogParams` is only provided to a component when it is opened in a modal. In other words, Angular's dependency injection doesn't know anything about a `ModalDialogParams` service unless the component is explicitly opened via NativeScript's `ModalService`, so this would break our ability to route to this component as we had originally set up, since Angular's DI would fail to recognize such a provider by default. In order to allow this component to continue working as a routing component, we will simply mark that argument as `@Optional()`, which will just set its value to null when not available instead of throwing a dependency injection error.

This will allow our component to be routed to, as well as be opened in a modal! Reuse in full swing!

In order to conditionally navigate to this component via routing, or open it in a modal, we can make a few small adjustments, bearing in mind that `RecorderModule` is lazy loaded, so we'll want to lazily load the module before opening it as a modal.

Open `app/modules/mixer/components/action-bar/action-bar.component.ts` and make the following modifications:

```
// angular
import { Component, Input, Output, EventEmitter } from '@angular/core';

// nativescript
import { RouterExtensions } from 'nativescript-angular/router';

import { PlayerService } from '../../../player/services/player.service';

@Component({
  moduleId: module.id,
  selector: 'action-bar',
  templateUrl: 'action-bar.component.html'
})
export class ActionBarComponent {
  ...
  @Output() showRecordModal: EventEmitter<any> = new EventEmitter();
  ...
  constructor(
```

```
    private router: RouterExtensions,
    private playerService: PlayerService
  ) { }
  public record() {
    if (this.playerService.composition &&
        this.playerService.composition.tracks.length) {
      // display recording UI as modal
      this.showRecordModal.next();
    } else {
      // navigate to it
      this.router.navigate(['/record']);
    }
  }
}
```

Here, we conditionally emit an event using `EventEmitter` with a Component `Output` decorator if the composition contains tracks; otherwise we navigate to the record view. We then adjust `Button` in the view template to use the method:

```
<ActionItem (tap)="record()" ios.position="right">
  <Button text="Record" class="action-item"></Button>
</ActionItem>
```

We can now modify `app/modules/mixer/components/mixer.component.html` to use `Output` by its name as a normal event:

```
<action-bar [title]="composition.name"
(showRecordModal)="showRecordModal()"></action-bar>
<GridLayout rows="*, auto" columns="*" class="page">
  <track-list [tracks]="composition.tracks" row="0" col="0"></track-list>
  <player-controls [composition]="composition" row="1" col="0"></player-
controls>
</GridLayout>
```

Now for the fun part. Since we'd love to be able to open any component in a modal, whether it's part of a lazy loaded module or not, let's add a new method to `DialogService` that can be used anywhere.

Make the following changes to `app/modules/core/services/dialog.service.ts`:

```
// angular
import { Injectable, NgModuleFactory, NgModuleFactoryLoader,
ViewContainerRef, NgModuleRef } from '@angular/core';

// nativescript
import * as dialogs from 'ui/dialogs';
import { ModalDialogService } from 'nativescript-
```

```
angular/directives/dialogs';

@Injectable()
export class DialogService {

  constructor(
    private moduleLoader: NgModuleFactoryLoader,
    private modalService: ModalDialogService
  ) { }

  public openModal(componentType: any, vcRef: ViewContainerRef, context?:
any, modulePath?: string): Promise<any> {
    return new Promise((resolve, reject) => {

      const launchModal = (moduleRef?: NgModuleRef<any>) => {
        this.modalService.showModal(componentType, {
          moduleRef,
          viewContainerRef: vcRef,
          context
        }).then(resolve, reject);
      };

      if (modulePath) {
        // lazy load module which contains component to open in modal
        this.moduleLoader.load(modulePath)
          .then((module: NgModuleFactory<any>) => {
            launchModal(module.create(vcRef.parentInjector));
          });
      } else {
        // open component in modal known to be available without lazy
loading
        launchModal();
      }
    });
  }
  ...
}
```

Here, we inject `ModalDialogService` and `NgModuleFactoryLoader` (which is actually `NSModuleFactoryLoader`, since, if you recall, we provided for in `Chapter 5`, *Routing and Lazy Loading*) to load any module on demand to open a Component (declared in that lazy loaded module) in a modal. *It also works for components that do not need to be lazy loaded*. In other words, it will optionally lazily load any module by its path, if provided, and then use its `NgModuleFactory` to get a module reference, which we can pass along as an option (via the `moduleRef` key) to `this.modalService.showModal` to open a Component declared in that lazily-loaded module.

This will come in handy again later; however, let's put it to use now by making the following changes to `app/modules/mixer/components/mixer.component.ts`:

```
// angular
import { Component, OnInit, OnDestroy, ViewContainerRef } from
'@angular/core';
import { ActivatedRoute } from '@angular/router';
import { Subscription } from 'rxjs/Subscription';

// app
import { DialogService } from '../../core/services/dialog.service';
import { MixerService } from '../services/mixer.service';
import { CompositionModel } from '../../shared/models';
import { RecordComponent } from
'../../recorder/components/record.component';

@Component({
 moduleId: module.id,
 selector: 'mixer',
 templateUrl: 'mixer.component.html'
})
export class MixerComponent implements OnInit, OnDestroy {

  public composition: CompositionModel;
  private _sub: Subscription;

  constructor(
    private route: ActivatedRoute,
    private mixerService: MixerService,
    private dialogService: DialogService,
    private vcRef: ViewContainerRef
  ) { }

  public showRecordModal() {
    this.dialogService.openModal(
      RecordComponent,
      this.vcRef,
      { isModal: true },
      './modules/recorder/recorder.module#RecorderModule'
    );
  }
  ...
}
```

This will lazily load `RecorderModule` and then open `RecordComponent` in a popup modal. Cool!

Finishing implementation with RecorderService

Now, let's finish this implementation with `RecorderService` in `app/modules/recorder/services/recorder.service.ts`:

```
// angular
import { Injectable } from '@angular/core';
import { Subject } from 'rxjs/Subject';
import { Subscription } from 'rxjs/Subscription';

// app
import { DialogService } from '../../core/services/dialog.service';
import { RecordModel, RecordState } from '../models';
import { PlayerService } from '../../player/services/player.service';
import { TrackModel } from '../../shared/models/track.model';

@Injectable()
export class RecorderService {
  public state$: Subject<number> = new Subject();
  public model: RecordModel;
  private _trackId: number;
  private _sub: Subscription;

  constructor(
    private playerService: PlayerService,
    private dialogService: DialogService
  ) { }

  public setupNewRecording() {
    this.model = new RecordModel();
    this._trackId = undefined; // reset

    this.model.on(this.model.events.stateChange,
this._stateHandler.bind(this));
    this._sub = this.playerService.complete$.subscribe(_ => {
      this.model.stopPlayback();
    });
  }

  public toggleRecord() {
    this.model.toggleRecord();
  }

  public togglePlay() {
    this.model.togglePlay();
  }

  public rewind() {
```

```
    this.playerService.seekTo(0); // reset to 0
  }

  public save() {
    this.model.save();
  }

  public cleanup() {
    // unbind event listener
    this.model.off(this.model.events.stateChange,
this._stateHandler.bind(this));
    this._sub.unsubscribe();

    if (!this.model.savedFilePath) {
      // user did not save recording, cleanup
      this.playerService.removeTrack(this._trackId);
    }
  }

  private _stateHandler(e) {
    this.state$.next(e.data);

    switch (e.data) {
      case RecordState.readyToRecord:
        this._stopMix();
        break;
      case RecordState.readyToPlay:
        this._stopMix();
        this._trackId = this.playerService
          .updateCompositionTrack(this._trackId, this.model.audioFilePath);
        break;
      case RecordState.playing:
        this._playMix();
        break;
      case RecordState.recording:
        this._playMix(this._trackId);
        break;
      case RecordState.saved:
        this._handleSaved();
        break;
    }
  }

  private _playMix(excludeTrackId?: number) {
    if (!this.playerService.playing) {
      // ensure mix plays
      this.playerService.togglePlay(excludeTrackId);
    }
```

```
  }

  private _stopMix() {
    if (this.playerService.playing) {
      // ensure mix stops
      this.playerService.togglePlay();
    }
    // always reset to beginning
    this.playerService.seekTo(0);
  }

  private _handleSaved() {
    this._sub.unsubscribe();
    this._stopMix();
    this.playerService
      .updateCompositionTrack(this._trackId, this.model.savedFilePath);
    this.playerService.saveComposition();
    this.model.finish();
  }
}
```

The pinnacle of our recording service is its ability to react to the model's state changes. This, in turn, emits an Observable stream notifying observers (our `RecordComponent`) when the state changes, as well as internally doing the work necessary to control `RecordModel` along with `PlayerService`. The critical key to our design is we want our active composition's tracks to play in the background while we record, so we can play along with the mix. This case is important:

```
case RecordState.readyToPlay:
  this._stopMix();
  this._trackId = this.playerService
    .updateCompositionTrack(this._trackId, this.model.audioFilePath);
  break;
```

When `RecordModel` is `readyToPlay`, we know that a recording has been created and is now ready to play. We stop the playing mix, get a reference to the recorded file's path. Then, we update `PlayerService` to queue up this new track to be played back. We will show the updated `PlayerService` in a moment, which handles adding the new file to the mix, but it adds a new `TrackPlayer` like everything else in our mix. However, the file points to a temporary recorded file at the moment, as we don't want to save the composition until the user decides to officially commit and save the recording. The recording session will allow the user to re-record again if they are not happy with the recording. This is why we hold a reference to `_trackId`. If a recording had already been added to the mix, we use that `_trackId` to exclude it when re-recording, since we would not want to hear back the recording we are re-recording over:

```
case RecordState.recording:
  this._playMix(this._trackId);
  break;
```

We also use it to clean up after ourselves if the user chose to cancel instead of saving:

```
public cleanup() {
  // unbind event listener
  this.model.off(this.model.events.stateChange,
this._stateHandler.bind(this));
  this._sub.unsubscribe();

  if (!this.model.savedFilePath) {
    // user did not save recording, cleanup
    this.playerService.removeTrack(this._trackId);
  }
}
```

Let's take a look at the modifications to `PlayerService` we need to make in order to support our recording:

```
...
import { MixerService } from '../../mixer/services/mixer.service';

@Injectable()
export class PlayerService {

  // default name of new tracks
  private _defaultTrackName: string = 'New Track';
  ...
  constructor(
    private ngZone: NgZone,
    private mixerService: MixerService
  ) { ... }
  ...
  public saveComposition() {
    this.mixerService.save(this.composition);
  }

  public togglePlay(excludeTrackId?: number) {
    if (this._trackPlayers.length) {
      this.playing = !this.playing;
      if (this.playing) {
        this.play(excludeTrackId);
      } else {
        this.pause();
      }
    }
```

```
  }
  public play(excludeTrackId?: number) {
    // for iOS playback sync
    let shortStartDelay = .01;
    let now = 0;

    for (let i = 0; i < this._trackPlayers.length; i++) {
      let track = this._trackPlayers[i];
      if (excludeTrackId !== track.trackId) {
        if (isIOS) {
          if (i == 0) now = track.player.ios.deviceCurrentTime;
          (<any>track.player).playAtTime(now + shortStartDelay);
        } else {
          track.player.play();
        }
      }
    }
  }

  public addTrack(track: ITrack): Promise<any> {
    return new Promise((resolve, reject) => {

      let trackPlayer = this._trackPlayers.find((p) => p.trackId ===
track.id);
      if (!trackPlayer) {
        // new track
        trackPlayer = new TrackPlayerModel();
        this._composition.tracks.push(track);
        this._trackPlayers.push(trackPlayer);
      } else {
        // update track
        this.updateTrack(track);
      }

      trackPlayer.load(
        track,
        this._trackComplete.bind(this),
        this._trackError.bind(this)
      ).then(_ => {
        // report longest duration as totalDuration
        this._updateTotalDuration();
        resolve();
      });
    })
  }

  public updateCompositionTrack(trackId: number, filepath: string): number
{
```

```
    let track;
    if (!trackId) {
      // Create a new track
      let cnt = this._defaultTrackNamesCnt();
      track = new TrackModel({
        name: `${this._defaultTrackName}${cnt ? ' ' + (cnt + 1) : ''}`,
        order: this.composition.tracks.length,
        filepath
      });
      trackId = track.id;
    } else {
      // find by id and update
      track = this.findTrack(trackId);
      track.filepath = filepath;
    }
    this.addTrack(track);
    return trackId;
  }

  private _defaultTrackNamesCnt() {
    return this.composition.tracks
      .filter(t => t.name.startsWith(this._defaultTrackName)).length;
  }
  ...
```

These changes will support our recorder's ability to interact with the active composition.

Note: Considerations when reusing a Component to lazy load in a modal as well as allow lazy loading via routing.

Angular services must be provided *only* at the *root* level *if they are intended to be singletons* shared across all lazy loaded modules, as well as the root module. `RecorderService` is lazy loaded with `RecordModule` when it is navigated to, as well as being opened in a modal. Since we are now injecting `PlayerService` into our `RecorderService` (which is lazily loaded) and `PlayerService` now injects `MixerService` (which is also lazily loaded as the root route in our app), we will have to create a problem where our services are no longer singletons. In fact, you may even see an error like this if you were to try and navigate to `RecordComponent`:

JS: ERROR Error: Uncaught (in promise): Error: No provider for PlayerService!

To solve this, we will drop the providers from `PlayerModule` and `MixerModule` (since those modules are both lazily loaded) and instead provide those services only in our `CoreModule`:

The modified `app/modules/player/player.module.ts` is as follows:

```
...
// import { PROVIDERS } from './services'; // commented out now

@NgModule({
  ...
  // providers: [...PROVIDERS], // no longer provided here
  ...
})
export class PlayerModule {}
```

The modified `app/modules/mixer/mixer.module.ts` is as follows:

```
...
// import { PROVIDERS } from './services'; // commented out now

@NgModule({
  ...
  // providers: [...PROVIDERS], // no longer provided here
  ...
})
export class MixerModule {}
```

Updated to provide these services as true singletons from `CoreModule` only, the code for `app/modules/core/core.module.ts` is as follows:

```
...
import { PROVIDERS } from './services';
import { PROVIDERS as MIXER_PROVIDERS } from '../mixer/services';
import { PROVIDERS as PLAYER_PROVIDERS } from '../player/services';

...

@NgModule({
  ...
  providers: [
    ...PROVIDERS,
    ...MIXER_PROVIDERS,
    ...PLAYER_PROVIDERS
  ],
  ...
})
export class CoreModule {
```

This is how you can solve these types of issues; but, this is exactly the reason why we recommend using Ngrx in `Chapter 10`, *@ngrx/store + @ngrx/effects for State Management*, coming up soon, as it can help alleviate these dependency injection issues.

At this point, our setup works nicely; but, it can be greatly improved and even simplified when we start integrating ngrx for a more Redux-style architecture. We have done a few reactive things here, such as our `RecordComponent` reacting to our service's `state$` observable; but, we needed to inject `MixerService` into `PlayerService`, which feels slightly wrong architecturally, since `PlayerModule` should not really have a dependency on anything `MixerModule` provides. Again, *this technically works just fine,* but when we start working with ngrx in `Chapter 10`, *@ngrx/store + @ngrx/effects for State Management*, you'll see how we can reduce our dependency mixing throughout the whole codebase.

Let's take a moment though, relax, and pat ourselves on the back, as this has been an impressive amount of work. Take a look at what the fruits of our labor are producing:

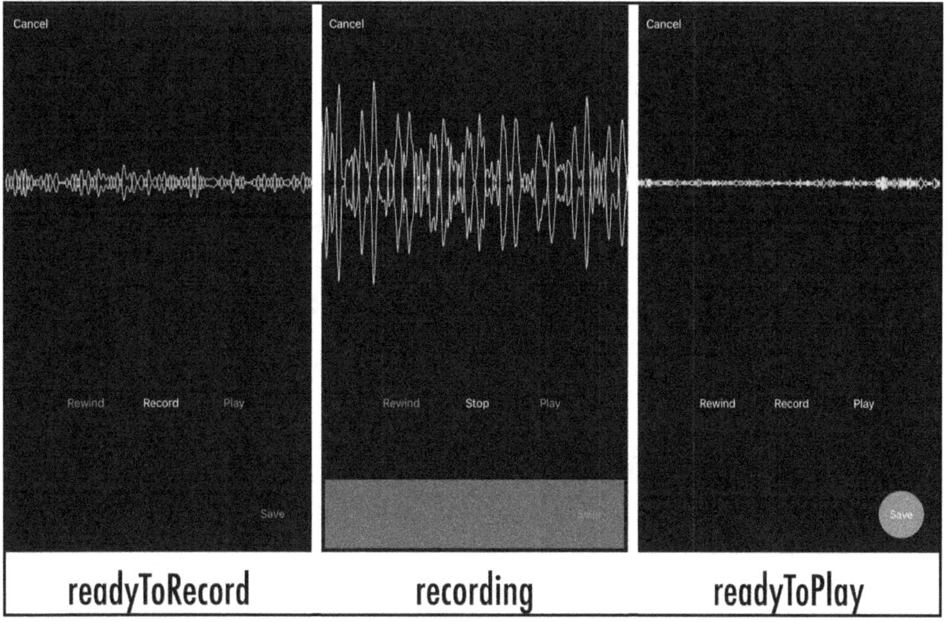

Phase 2 – Building an audio recorder for Android

Believe it or not we've actually done most of the heavy lifting to make this work on Android already! That's the beauty of NativeScript. Designing an API that makes sense, as well as an architecture that can plug/play underlying native APIs, is key to NativeScript development. At this point, we just need to plug in the Android pieces into the shape we have designed. So, to summarize, we now have the following:

- `RecorderService` that works in tandem with `PlayerService` to coordinate our multitrack handling abilities
- A Waveform view that is flexible and ready to provide an Android implementation under the hood
- `RecordModel` that should tap into the appropriate underlying target platform APIs and be ready for Android details to be plugged into
- Built interfaces defining the shape of the model, for Android models to simply implement to know which API they should define

Let's get to work.

We want to rename `record.model.ts` to `record.model.ios.ts`, since it's specific to iOS, but before doing so, we will want a TypeScript definition file (`.d.ts`) for it, so our codebase can continue importing as `'record.model'`. There are several ways this could be done, including just manually writing one out. However, the tsc compiler has a handy `-d` flag, which will generate definition files for us:

```
tsc app/modules/recorder/models/record.model.ts references.d.ts -d true
```

This will spit out a ton of TypeScript warnings and errors; but, it doesn't matter in this case, since our definition file will be generated correctly. We don't need to generate JavaScript, just the definition, so you can ignore the wall of issues that results.

We now have two new files:

- `record-common.model.d.ts` (*you can delete this as we won't need it*)
- `record.model.d.ts`

The `record-common.model` file is imported by `RecordModel`, which is why a definition was generated for it as well; but, you can *delete* that. Now, we have the definition file, but we want to modify it slightly. We don't need any of the `private` declarations and/or any native types it included; you would notice it contained the following:

```
...
readonly target: AKMicrophone;
readonly recorder: AKNodeRecorder;
...
```

Since those are iOS-specific, we'll want to type those as *any*, so it's applicable to both iOS and Android. This is what things look like with our modifications:

```
import { Observable } from 'data/observable';
import { IRecordModel, IRecordEvents } from './common';
export declare class RecordModel extends Observable implements IRecordModel
{
  readonly events: IRecordEvents;
  readonly target: any;
  readonly recorder: any;
  readonly audioFilePath: string;
  state: number;
  savedFilePath: string;
  toggleRecord(): void;
  togglePlay(): void;
  stopPlayback(): void;
  save(): void;
  dispose(): void;
  finish(): void;
}
```

Perfect, now rename `record.model.ts` to `record.model.ios.ts`. We have now finalized our iOS implementation, as well as ensured maximum code reuse to turn our focus to Android. NativeScript will use the target platform suffix files at build time, so you don't ever need to worry that iOS-only code would end up on Android and vice versa.

The `.d.ts` definition file we generated previously will be used at JavaScript transpilation time by the TypeScript compiler, whereas the runtime will use the platform-specific JS files (without the extension).

Okay, now create `app/modules/recorder/models/record.model.android.ts`:

```
import { Observable } from 'data/observable';
import { IRecordModel, IRecordEvents, RecordState, documentsFilePath } from
'./common';

export class RecordModel extends Observable implements IRecordModel {

  // available events to listen to
  private _events: IRecordEvents;

  // recorder
  private _recorder: any;

  // state
  private _state: number = RecordState.readyToRecord;

  // the final saved path to use
  private _savedFilePath: string;

  constructor() {
    super();
    this._setupEvents();
    // TODO
  }

  public get events(): IRecordEvents {
    return this._events;
  }

  public get target() {
    // TODO
  }

  public get recorder(): any {
    return this._recorder;
  }

  public get audioFilePath(): string {
    return ''; // TODO
  }

  public get state(): number {
    return this._state;
  }

  public set state(value: number) {
    this._state = value;
```

```
  this._emitEvent(this._events.stateChange, this._state);
}

public get savedFilePath() {
  return this._savedFilePath;
}

public set savedFilePath(value: string) {
  this._savedFilePath = value;
  if (this._savedFilePath)
    this.state = RecordState.saved;
}

public toggleRecord() {
  if (this._state !== RecordState.recording) {
    // just force ready to record
    // when coming from any state other than recording
    this.state = RecordState.readyToRecord;
  }

  switch (this._state) {
    case RecordState.readyToRecord:
      this.state = RecordState.recording;
      break;
    case RecordState.recording:
      this._recorder.stop();
      this.state = RecordState.readyToPlay;
      break;
  }
}

public togglePlay() {
  if (this._state === RecordState.readyToPlay) {
    this.state = RecordState.playing;
  } else {
    this.stopPlayback();
  }
}

public stopPlayback() {
  if (this.state !== RecordState.recording) {
    this.state = RecordState.readyToPlay;
  }
}

public save() {
  // we will want to do this
  // this.savedFilePath = documentsFilePath(fileName);
```

```
  }

  public dispose() {
    // TODO
  }

  public finish() {
    this.state = RecordState.finish;
  }

  private _emitEvent(eventName: string, data?: any) {
    let event = {
      eventName,
      data,
      object: this
    };
    this.notify(event);
  }

  private _setupEvents() {
    this._events = {
      stateChange: 'stateChange'
    };
  }
}
```

This may look a whole lot like the iOS side, and that's because it will be nearly the same! In fact, this setup works great, so now we just want to fill in the Android specifics.

Using nativescript-audio's TNSRecorder for Android in our RecordModel

We could use some fancy Android APIs and/or libraries for our recorder, but in this case, the **nativescript-audio** plugin we're using for our cross-platform multitrack player also provides a cross-platform recorder. We could have even used it with iOS, but we wanted to specifically work with AudioKit's powerful APIs there. However, here on Android, let's use the recorder from the plugin and make the following modifications to `record.model.android.ts`:

```
import { Observable } from 'data/observable';
import { IRecordModel, IRecordEvents, RecordState, documentsFilePath } from
'./common';
import { TNSRecorder, AudioRecorderOptions } from 'nativescript-audio';
import { Subject } from 'rxjs/Subject';
```

```
import * as permissions from 'nativescript-permissions';

declare var android: any;
const RECORD_AUDIO = android.Manifest.permission.RECORD_AUDIO;

export class RecordModel extends Observable implements IRecordModel {

  // available events to listen to
  private _events: IRecordEvents;

  // target as an Observable
  private _target$: Subject<number>;

  // recorder
  private _recorder: TNSRecorder;
  // recorder options
  private _options: AudioRecorderOptions;
  // recorder mix meter handling
  private _meterInterval: number;

  // state
  private _state: number = RecordState.readyToRecord;

  // tmp file path
  private _filePath: string;
  // the final saved path to use
  private _savedFilePath: string;

  constructor() {
    super();
    this._setupEvents();
    // prepare Observable as our target
    this._target$ = new Subject();

    // create recorder
    this._recorder = new TNSRecorder();
    this._filePath = documentsFilePath(`recording-${Date.now()}.m4a`);
    this._options = {
      filename: this._filePath,
      format: android.media.MediaRecorder.OutputFormat.MPEG_4,
      encoder: android.media.MediaRecorder.AudioEncoder.AAC,
      metering: true, // critical to feed our waveform view
      infoCallback: (infoObject) => {
        // just log for now
        console.log(JSON.stringify(infoObject));
      },
      errorCallback: (errorObject) => {
        console.log(JSON.stringify(errorObject));
```

```
      }
    };
  }

  public get events(): IRecordEvents {
    return this._events;
  }

  public get target() {
    return this._target$;
  }

  public get recorder(): any {
    return this._recorder;
  }

  public get audioFilePath(): string {
    return this._filePath;
  }

  public get state(): number {
    return this._state;
  }

  public set state(value: number) {
    this._state = value;
    this._emitEvent(this._events.stateChange, this._state);
  }

  public get savedFilePath() {
    return this._savedFilePath;
  }

  public set savedFilePath(value: string) {
    this._savedFilePath = value;
    if (this._savedFilePath)
      this.state = RecordState.saved;
  }

  public toggleRecord() {
    if (this._state !== RecordState.recording) {
      // just force ready to record
      // when coming from any state other than recording
      this.state = RecordState.readyToRecord;
    }

    switch (this._state) {
      case RecordState.readyToRecord:
```

```
      if (this._hasPermission()) {
        this._recorder.start(this._options).then((result) => {
          this.state = RecordState.recording;
          this._initMeter();
        }, (err) => {
          this._resetMeter();
        });
      } else {
        permissions.requestPermission(RECORD_AUDIO).then(() => {
          // simply engage again
          this.toggleRecord();
        }, (err) => {
          console.log('permissions error:', err);
        });
      }
      break;
    case RecordState.recording:
      this._resetMeter();
      this._recorder.stop();
      this.state = RecordState.readyToPlay;
      break;
  }
}

public togglePlay() {
  if (this._state === RecordState.readyToPlay) {
    this.state = RecordState.playing;
  } else {
    this.stopPlayback();
  }
}

public stopPlayback() {
  if (this.state !== RecordState.recording) {
    this.state = RecordState.readyToPlay;
  }
}

public save() {
  // With Android, filePath will be the same, just make it final
  this.savedFilePath = this._filePath;
}

public dispose() {
  if (this.state === RecordState.recording) {
    this._recorder.stop();
  }
  this._recorder.dispose();
```

```
  }

  public finish() {
    this.state = RecordState.finish;
  }

  private _initMeter() {
    this._resetMeter();
    this._meterInterval = setInterval(() => {
      let meters = this.recorder.getMeters();
      this._target$.next(meters);
    }, 200); // use 50 for production - perf is better on devices
  }

  private _resetMeter() {
    if (this._meterInterval) {
      clearInterval(this._meterInterval);
      this._meterInterval = undefined;
    }
  }

  private _hasPermission() {
    return permissions.hasPermission(RECORD_AUDIO);
  }

  private _emitEvent(eventName: string, data?: any) {
    let event = {
      eventName,
      data,
      object: this
    };
    this.notify(event);
  }

  private _setupEvents() {
    this._events = {
      stateChange: 'stateChange'
    };
  }
}
```

Wow! Okay, a lot of interesting things going on here. Let's get one necessary thing out of the way for Android and ensure for API level 23+ that permissions are properly handled. For this, you can install the permissions plugin:

```
tns plugin add nativescript-permissions
```

We also want to ensure our manifest file contains the proper permission key.

Open `app/App_Resources/Android/AndroidManifest.xml` and add the following in the correct place:

```
<uses-permission android:name="android.permission.READ_EXTERNAL_STORAGE"/>
<uses-permission android:name="android.permission.WRITE_EXTERNAL_STORAGE"/>
<uses-permission android:name="android.permission.INTERNET"/>
<uses-permission android:name="android.permission.RECORD_AUDIO"/>
```

We use the nativescript-audio plugin's `TNSRecorder` as our implementation and wire things up accordingly to its API. `AudioRecorderOptions` provides a `metering` option, allowing the ability to monitor the microphone's meters via an interval.

What is most versatile about our overall design is that our model's target can literally be anything. In this case, we create a RxJS Subject observable as `_target$`, which is then returned as our target getter. This allows us to emit the microphone's meter value through the `Subject` observable for consumption by our Waveform. You will see in a moment how we will take advantage of this.

We are now ready to move on to our Waveform implementation for Android.

Just like we did for the model, we will want to refactor the common bits into a shared file and handle the suffix.

Create `app/modules/shared/native/waveform-common.ts`:

```
import { View } from 'ui/core/view';

export type WaveformType = 'mic' | 'file';

export interface IWaveformModel {
  readonly target: any;
  dispose(): void;
}

export interface IWaveform extends View {
  type: WaveformType;
  model: IWaveformModel;
  createNativeView(): any;
  initNativeView(): void;
  disposeNativeView(): void;
}
```

Then, just adjust app/modules/shared/native/waveform.ts to use it:

```
...
import { IWaveform, IWaveformModel, WaveformType } from './waveform-
common';

export class Waveform extends View implements IWaveform {
  ...
```

Before renaming our waveform to contain an .ios suffix, let's generate a TypeScript definition file for it first:

```
tsc app/modules/shared/native/waveform.ts references.d.ts -d true --lib
es6,dom,es2015.iterable --target es5
```

You may again see TypeScript errors or warnings, but we don't need to worry about those, as it should have still generated a waveform.d.ts file. Let's simplify it slightly to contain only the parts that are applicable to both iOS and Android:

```
import { View } from 'ui/core/view';
export declare type WaveformType = 'mic' | 'file';
export interface IWaveformModel {
  readonly target: any;
  dispose(): void;
}
export interface IWaveform extends View {
  type: WaveformType;
  model: IWaveformModel;
  createNativeView(): any;
  initNativeView(): void;
  disposeNativeView(): void;
}
export declare class Waveform extends View implements IWaveform {}
```

Okay, now, rename waveform.ts to waveform.ios.ts and create app/modules/shared/native/waveform.android.ts:

```
import { View } from 'ui/core/view';
import { Color } from 'color';
import { IWaveform, IWaveformModel, WaveformType } from './common';

export class Waveform extends View implements IWaveform {
  private _model: IWaveformModel;
  private _type: WaveformType;

  public set type(value: WaveformType) {
    this._type = value;
  }
```

```
    public get type() {
      return this._type;
    }

    public set model(value: IWaveformModel) {
      this._model = value;
    }

    public get model() {
      return this._model;
    }

    createNativeView() {
      switch (this.type) {
        case 'mic':
          // TODO: this.nativeView = ?
          break;
        case 'file':
          // TODO: this.nativeView = ?
          break;
      }
      return this.nativeView;
    }

    initNativeView() {
      // TODO
    }

    disposeNativeView() {
      if (this.model && this.model.dispose) this.model.dispose();
    }
  }
```

Okay, excellent! This is the barebones setup we will need, *but what native Android view should we use?*

If you're looking around for open source Android libs, you may come across a group of very talented developers with **Yalantis**, a fantastic mobile development company based out of Ukraine. Roman Kozlov and his team created an open source project, **Horizon**, which provides beautiful audio visualizations:
`https://github.com/Yalantis/Horizon`
`https://yalantis.com/blog/horizon-open-source-library-for-sound-visualization/`

Just like for iOS, we also want to prepare for a multifaceted Waveform view that can also render a static waveform for just a file. Looking further through the open source options, we may come across another wonderfully talented team with **Semantive**, based in Warsaw, the sprawling capital of Poland. They created an incredibly powerful Waveform view for Android:
`https://github.com/Semantive/waveform-android`

Let's integrate both of these libraries for our Android Waveform integration.

Similar to how we integrated AudioKit for iOS, let's create a folder in the root called `android-waveform-libs` with the following setup, that provides `include.gradle`:

```
android {
  productFlavors {
    "android-waveform-libs" {
      dimension "android-waveform-libs"
    }
  }
}

repositories {
  maven { url "https://jitpack.io" }
}

dependencies {
  compile "com.yalantis:eqwaves:1.0.1"
  compile "com.github.Semantive:waveform-android:v1.2"
}
```

Why deviate from the `nativescript-` prefix when including native libs? The prefix is a good way to go if you plan to refactor the internal plugin into an open source plugin published via npm for the community down the road, using `https://github.com/NathanWalker/nativescript-plugin-seed` for instance.

Sometimes, you just need to integrate several native libs for a specific platform, as we are in this case, so we don't really need the `nativescript-` prefix on our folder.

We make sure to add `package.json`, so we can add these native libs like we would any other plugin:

```
{
  "name": "android-waveform-libs",
  "version": "1.0.0",
  "nativescript": {
    "platforms": {
      "android": "3.0.0"
    }
  }
}
```

Now, we simply add them as a plugin to our project:

```
tns plugin add android-waveform-libs
```

We are now ready to integrate these libs into our Waveform view. Let's make the following modifications to the `app/modules/shared/native/waveform.android.ts` file:

```
import { View } from 'ui/core/view';
import { Color } from 'color';
import { Subscription } from 'rxjs/Subscription';
import { IWaveform, IWaveformModel, WaveformType } from './common';
import { screen } from 'platform';

declare var com;
declare var android;
const GLSurfaceView = android.opengl.GLSurfaceView;
const AudioRecord = android.media.AudioRecord;

// Horizon recorder waveform
// https://github.com/Yalantis/Horizon
const Horizon = com.yalantis.waves.util.Horizon;
// various recorder settings
const RECORDER_SAMPLE_RATE = 44100;
const RECORDER_CHANNELS = 1;
const RECORDER_ENCODING_BIT = 16;
const RECORDER_AUDIO_ENCODING = 3;
const MAX_DECIBELS = 120;

// Semantive waveform for files
// https://github.com/Semantive/waveform-android
const WaveformView =
com.semantive.waveformandroid.waveform.view.WaveformView;
const CheapSoundFile =
com.semantive.waveformandroid.waveform.soundfile.CheapSoundFile;
```

```
const ProgressListener =
com.semantive.waveformandroid.waveform.soundfile.CheapSoundFile.ProgressLis
tener;

export class Waveform extends View implements IWaveform {
  private _model: IWaveformModel;
  private _type: WaveformType;
  private _initialized: boolean;
  private _horizon: any;
  private _javaByteArray: Array<any>;
  private _waveformFileView: any;
  private _sub: Subscription;

  public set type(value: WaveformType) {
    this._type = value;
  }

  public get type() {
    return this._type;
  }

  public set model(value: IWaveformModel) {
    this._model = value;
    this._initView();
  }

  public get model() {
    return this._model;
  }

  createNativeView() {
    switch (this.type) {
      case 'mic':
        this.nativeView = new GLSurfaceView(this._context);
        this.height = 200; // GL view needs height
        break;
      case 'file':
        this.nativeView = new WaveformView(this._context, null);
        this.nativeView.setSegments(null);
        this.nativeView.recomputeHeights(screen.mainScreen.scale);

        // disable zooming and touch events
        this.nativeView.mNumZoomLevels = 0;
        this.nativeView.onTouchEvent = function (e) { return false; }
        break;
    }
    return this.nativeView;
  }
```

```
  initNativeView() {
    this._initView();
  }

  disposeNativeView() {
    if (this.model && this.model.dispose) this.model.dispose();
    if (this._sub) this._sub.unsubscribe();
  }

  private _initView() {
    if (!this._initialized && this.nativeView && this.model) {
      if (this.type === 'mic') {
        this._initialized = true;
        this._horizon = new Horizon(
          this.nativeView,
          new Color('#000').android,
          RECORDER_SAMPLE_RATE,
          RECORDER_CHANNELS,
          RECORDER_ENCODING_BIT
        );

        this._horizon.setMaxVolumeDb(MAX_DECIBELS);
        let bufferSize = 2 * AudioRecord.getMinBufferSize(
          RECORDER_SAMPLE_RATE, RECORDER_CHANNELS,
RECORDER_AUDIO_ENCODING);
        this._javaByteArray = Array.create('byte', bufferSize);

        this._sub = this._model.target.subscribe((value) => {
          this._javaByteArray[0] = value;
          this._horizon.updateView(this._javaByteArray);
        });
      } else {
        let soundFile = CheapSoundFile.create(this._model.target,
          new ProgressListener({
            reportProgress: (fractionComplete: number) => {
              console.log('fractionComplete:', fractionComplete);
              return true;
            }
          }));

        setTimeout(() => {
          this.nativeView.setSoundFile(soundFile);
          this.nativeView.invalidate();
        }, 0);
      }
    }
  }
}
```

We begin our Android implementation by defining the `const` references to the various packaged classes we need to access, to alleviate having to reference the fully qualified package location each time in our Waveform. Just like on the iOS side, we design a dual-purpose Waveform by allowing the type (`'mic'` or `'file'`) to drive which rendering to use. This allows us to reuse this with our record view for real-time microphone visualization and the other to statically render our tracks as Waveforms (more on that soon!).

The Horizon lib utilizes Android's `GLSurfaceView` as the primary rendering, hence:

```
this.nativeView = new GLSurfaceView(this._context);
this.height = 200; // GL view needs height
```

During development, we found that `GLSurfaceView` requires at least a height to constrain it, otherwise it would render at full screen height. Therefore, we explicitly set a reasonable `height` of `200` to the custom NativeScript view, which will automatically handle measuring the native view for us. Interestingly, we also found that sometimes our model setter would fire *before* `initNativeView` and other times *after*. Because the model is a critical binding for initializing our Horizon view, we designed a custom internal `_initView` method with the appropriate conditional, which could be called from `initNativeView`, as well as after our model setter fired. The condition (`!this._initialized && this.nativeView && this.model`) ensures it's only ever initialized once though. This is the way to handle any potential race conditions around the sequence of these method calls.

The native `Horizon.java` class provides an `update` method that expects a Java byte array with a signature:

```
updateView(byte[] buffer)
```

What we do in NativeScript for this is retain a reference to a construct that will represent this native Java byte array with the following:

```
let bufferSize = 2 * AudioRecord.getMinBufferSize(
  RECORDER_SAMPLE_RATE, RECORDER_CHANNELS, RECORDER_AUDIO_ENCODING);
this._javaByteArray = Array.create('byte', bufferSize);
```

Utilizing Android's `android.media.AudioRecord` class, in conjunction with the various recorder settings that we set up, we are able to gather an initial `bufferSize`, that we use to initialize our byte array size.

We then take advantage of our overall versatile design, wherein our model's target in this implementation is an rxjs Subject Observable, allowing us to subscribe to its event stream. For the `'mic'` type, this stream will be the metering value changes from the recorder, which we use to fill our byte array and in turn update the `Horizon` view:

```
this._sub = this._model.target.subscribe((value) => {
  this._javaByteArray[0] = value;
  this._horizon.updateView(this._javaByteArray);
});
```

This provides our recorder a nice visualization, which will animate as the input level changes. Here's a preview; however, the style is still a little ugly, since we have not applied any CSS polish just yet:

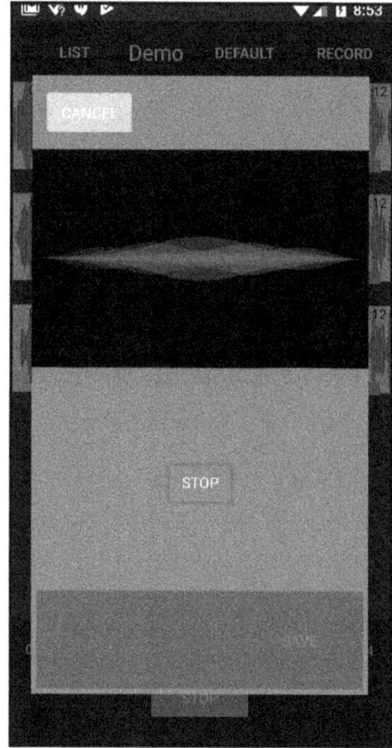

For our static audio file waveform rendering, we initialize `WaveformView` with the Android context. We then use its API to configure it for our use during construction in `createNativeView`.

During initialization, we create an instance of `CheapSoundFile` as required by `WaveformView`, and interestingly, we use `setSoundFile` inside `setTimeout`, alongside a call to `this.nativeView.invalidate()`, which is calling invalidate on `WaveformView`. This causes the native view to update with the processed file, as follows (again, we will address the styling polish later):

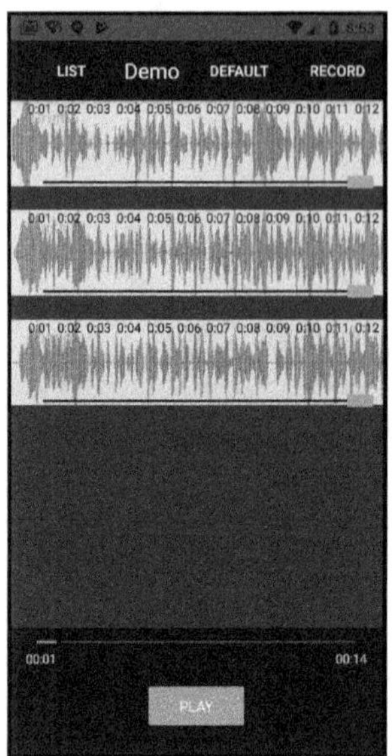

Summary

This chapter has introduced a wealth of powerful concepts and techniques on how to work with native APIs on iOS and Android. Knowing how to work with open source native libraries is fundamental to getting the most out of your app development and achieving the feature set you are after. Direct access to these APIs right from TypeScript gives you the luxury of never leaving your preferred development environment, as well as engaging with the languages you love in a fun and accessible way.

Additionally, learning solid practices around how/when to create custom NativeScript views and interworking them throughout your Angular app are among the key elements to leverage the most of this tech stack.

In the next chapter, we will provide some extra goodies by empowering our track list view with more bells and whistles, leveraging some of what you've learned here.

9
Empowering Your Views

The combination of Angular and NativeScript is fun to develop with and powerful beyond measure for mobile development. Whether you need to provide services to engage with a mobile device's hardware capabilities, such as audio recording or enrich your app's usability with engaging views, NativeScript for Angular provides exciting opportunities.

Let's continue with several concepts we developed in the preceding chapter to provide an alternate rich view of our tracks while reusing everything we've covered so far, in addition to a few new tips/tricks.

In this chapter, we will cover the following topics:

- Using multiple item row templates with `ListView` and `templateSelector`
- Handling row template changes with `ListView` and when/how to refresh them
- Using `ngModel` data binding via `NativeScriptFormsModule`
- Leveraging a shared singleton service for sharing state across multiple modules
- Serializing data before storing and hydrating upon retrieval from a persisted state
- Leveraging and reusing Angular directives to enrich the NativeScript Slider with more unique characteristics

Multiple item templates with NativeScript's ListView

All throughout `Chapter 8`, *Building an audio recorder*, we designed a dual-purpose custom NativeScript Waveform view, which taps into various native libraries for iOS and Android, specifically to enrich our composition's track listing view. Let's proceed by reusing our versatile Waveform view for our track listing view. This will also give us a way to display mixing slider controls (often referred to in audio production and sound engineering as a Fader) alongside our tracks to allow the user to mix each track's volume level in the overall composition. Let's set up our `ListView` of `TrackListComponent` with the ability to provide the user with two different ways to view and work with their tracks, each with their own unique utility. While we're at it, we'll also take this opportunity to finally wire up the `mute` switch on our tracks.

Let's make the following modifications to `app/modules/player/components/track-list/track-list.component.html`:

```
<ListView #listview [items]="tracks | orderBy: 'order'" class="list-group"
  [itemTemplateSelector]="templateSelector">
  <ng-template let-track="item" nsTemplateKey="default">
    <GridLayout rows="auto" columns="100,*,100" class="list-group-item"
      [class.muted]="track.mute">
      <Button text="Record" (tap)="record(track)" row="0" col="0" class="c-
ruby"></Button>
      <Label [text]="track.name" row="0" col="1" class="h2"></Label>
      <Switch row="0" col="2" class="switch"
[(ngModel)]="track.mute"></Switch>
    </GridLayout>
  </ng-template>

  <ng-template let-track="item" nsTemplateKey="waveform">
    <AbsoluteLayout [class.muted]="track.mute">
      <Waveform class="waveform w-full" top="0" left="0" height="80"
        [model]="track.model"
        type="file"
        plotColor="#888703"
        fill="true"
        mirror="true"
        plotType="buffer"></Waveform>

      <Label [text]="track.name" row="0" col="1" class="h3 track-name-
float"
        top="5" left="20"></Label>
      <Slider slim-slider="fader.png" minValue="0" maxValue="1"
```

```
        width="94%" top="50" left="0"
        [(ngModel)]="track.volume" class="slider fader"></Slider>
    </AbsoluteLayout>
  </ng-template>
</ListView>
```

There's a lot of interesting things happening here. First of all, `[itemTemplateSelector]="templateSelector"` provides the ability to change our `ListView` item rows on the fly. The result of the `templateSelector` function should be a string, which matches the value provided on any ng-template's `ngTemplateKey` attribute. To make all this work, we will need several things in place, starting with the `Component` that has access to the `ListView` via `#listview` and `ViewChild`:

```
// angular
import { Component, Input, ViewChild, ElementRef } from '@angular/core';
import { Router } from '@angular/router';

// nativescript
import { ListView } from 'ui/list-view';

// app
import { ITrack } from '../../../shared/models';
import { AuthService, DialogService } from '../../../core/services';
import { PlayerService } from '../../services/player.service';

@Component({
 moduleId: module.id,
 selector: 'track-list',
 templateUrl: 'track-list.component.html',
})
export class TrackListComponent {

  public templateSelector: Function;
  @Input() tracks: Array<ITrack>;
  @ViewChild('listview') _listviewRef: ElementRef;
  private _listview: ListView;
  private _sub: any;

  constructor(
    private authService: AuthService,
    private dialogService: DialogService,
    private router: Router,
    private playerService: PlayerService
  ) {
    this.templateSelector = this._templateSelector.bind(this);
  }
```

```
ngOnInit() {
  this._sub = this.playerService.trackListViewChange$.subscribe(() => {
    // since this involves our templateSelector, ensure ListView knows
about it
    // refresh list
    this._listview.refresh();
  });
}

ngAfterViewInit() {
  this._listview = <ListView>this._listviewRef.nativeElement;
}

private _templateSelector(item: ITrack, index: number, items: ITrack[]) {
  return this.playerService.trackListViewType;
}
...
```

We set up a `ViewChild` to retain a reference to our `ListView` which we will use later to call `this._listview.refresh()` . This is required in Angular when we need the `ListView` to update the display after changes. The first surprise will likely be the injection of `PlayerService`, and the second might be the `this.templateSelector = this._templateSelector.bind(this)`. The `templateSelector` binding is not scope bound, and since we need it to return a property reference from our `this.playerService` on the `Component`, we ensure that it is properly bound to the scope of the `Component` by binding a `Function` reference. We will use `PlayerService` as a conduit at this point to help communicate the state from the `ActionBarComponent` that lives in the `MixerModule`.

This example shows how services can help communicate the state throughout your app. However, this practice can be greatly improved by utilizing `ngrx` to help lessen interwoven dependencies and unlock a purely reactive setup with Redux-style architecture. @ngrx enhancements will be covered in `Chapter 10`, *@ngrx/store + @ngrx/effects for State Management*.

Our **View Toggle** button will be in the `ActionBar` (in the `MixerModule`), and we will want to tap there to switch our `ListView`, which lives inside our `PlayerModule`. The `PlayerService` is a singleton at the moment (provided by `CoreModule`) and is shared across the entire app, so it's a perfect candidate to aid here. Let's take a look at our `ActionBarComponent` changes first in `app/modules/mixer/components/action-bar/action-bar.component.ios.html`:

```
<ActionBar [title]="title" class="action-bar">
  <ActionItem nsRouterLink="/mixer/home">
```

```
      <Button text="List" class="action-item"></Button>
    </ActionItem>
    <ActionItem (tap)="toggleList()" ios.position="right">
      <Button [text]="toggleListText" class="action-item"></Button>
    </ActionItem>
    <ActionItem (tap)="record()" ios.position="right">
      <Button text="Record" class="action-item"></Button>
    </ActionItem>
  </ActionBar>
```

Then, we'll take a look at the changes in `app/modules/mixer/components/action-bar/action-bar.component.android.html`:

```
<ActionBar class="action-bar">
  <GridLayout rows="auto" columns="auto,*,auto,auto" class="action-bar">
    <Button text="List" nsRouterLink="/mixer/home"
      class="action-item" row="0" col="0"></Button>
    <Label [text]="title" class="action-bar-title text-center" row="0"
col="1"></Label>
    <Button [text]="toggleListText" (tap)="toggleList()"
      class="action-item" row="0" col="2"></Button>
    <Button text="Record" (tap)="record()"
      class="action-item" row="0" col="3"></Button>
  </GridLayout>
</ActionBar>
```

We'll also take a look at the changes in the `Component`:

```
...
import { PlayerService } from '../../../player/services/player.service';

@Component({
  moduleId: module.id,
  selector: 'action-bar',
  templateUrl: 'action-bar.component.html'
})
export class ActionBarComponent {
  ...
  public toggleListText: string = 'Waveform';

  constructor(
    private router: RouterExtensions,
    private playerService: PlayerService
  ) { }
  ...
  public toggleList() {
    // later we can use icons, using labels for now
    let type = this.playerService.trackListViewType === 'default' ?
```

```
'waveform' : 'default';
    this.playerService.trackListViewType = type;
    this.toggleListText = type === 'default' ? 'Waveform' : 'Default';
  }
}
```

As you can see, we added a button to the `ActionBar`, which will use the label `Waveform` or `Default`, depending on its state. Then, we used `PlayerService` to modify a new setter, **`this.playerService.trackListViewType = type`**. Let's take a look at `app/modules/player/services/player.service.ts` now:

```
...
@Injectable()
export class PlayerService {
  ...
  // communicate state changes from ActionBar to anything else
  public trackListViewChange$: Subject<string> = new Subject();
  ...
  public get trackListViewType() {
    return this._trackListViewType;
  }

  public set trackListViewType(value: string) {
    this._trackListViewType = value;
    this.trackListViewChange$.next(value);
  }
  ...
```

This gets the job done.

As mentioned, we will improve this setup in the next chapter with ngrx, which is all about polishing and simplifying the way we handle our app's state.

There are a couple more things that we will need to do to ensure that all our new additions work. For starters, the `[(ngModel)]` bindings will *not* work at all without the `NativeScriptFormsModule`.

If you use the `ngModel` bindings in your component's view, you must ensure that the module that declares your `Component` imports the `NativeScriptFormsModule`. If it uses a `SharedModule`, ensure that the `SharedModule` imports and exports the `NativeScriptFormsModule`.

Let's add the module mentioned in the preceding tip to our `SharedModule` so that all of our modules can use `ngModel` wherever needed:

```
...
import { NativeScriptFormsModule } from 'nativescript-angular/forms';
...
@NgModule({
  imports: [
    NativeScriptModule,
    NativeScriptRouterModule,
    NativeScriptFormsModule
  ],
  ...
  exports: [
    NativeScriptModule,
    NativeScriptRouterModule,
    NativeScriptFormsModule,
    ...PIPES
  ]
})
export class SharedModule {}
```

We will now need each track's mute and volume property changes to notify our audio player. This involves changing our `TrackModel` slightly to account for this new functionality; to do that, open `app/modules/shared/models/track.model.ts`:

```
import { BehaviorSubject } from 'rxjs/BehaviorSubject';
...
export class TrackModel implements ITrack {
  public id: number;
  public filepath: string;
  public name: string;
  public order: number;
  public model: any;

  public volume$: BehaviorSubject<number>;

  private _volume: number = 1; // default full volume
  private _mute: boolean;
  private _origVolume: number; // return to after unmute

  constructor(model?: ITrack) {
    this.volume$ = new BehaviorSubject(this._volume);
    ...
  }

  public set mute(value: boolean) {
```

```
    this._mute = value;
    if (this._mute) {
      this._origVolume = this._volume;
      this.volume = 0;
    } else {
      this.volume = this._origVolume;
    }
  }

  public get mute() {
    return this._mute;
  }

  public set volume(value: number) {
    this._volume = value;
    this.volume$.next(this._volume);
    if (this._volume > 0 && this._mute) {
      // if just increasing volume from a muted state
      // ensure it's unmuted
      this._origVolume = this._volume;
      this._mute = false;
    }
  }

  public get volume() {
    return this._volume;
  }
}
```

We now want to modify our `TrackPlayerModel` to work in tandem with these new features. Earlier, we used to retain just the `trackId`; however, with this new addition, it would be helpful to keep a reference to the entire `TrackModel` object, so open `app/modules/shared/models/track-player.model.ts` and make the following changes:

```
...
import { Subscription } from 'rxjs/Subscription';
...
interface ITrackPlayer {
  track: TrackModel; // was trackId only
  duration: number;
  readonly player: TNSPlayer;
}
...
export class TrackPlayerModel implements ITrackPlayer {
  public track: TrackModel;
  ...
```

```
  private _sub: Subscription;
  ...
  public load(track: TrackModel, complete: Function, error: Function):
Promise<number> {
    return new Promise((resolve, reject) => {
      this.track = track;

      this._player.initFromFile({
        ...
      }).then(() => {
        ...
        // if reloading track, clear subscription before subscribing again
        if (this._sub) this._sub.unsubscribe();
        this._sub = this.track.volume$.subscribe((value) => {
          if (this._player) {
            // react to track model property changes
            this._player.volume = value;
          }
        });
      }, reject);
    });
  }
  ...
  public cleanup() {
    // cleanup and dispose player
    if (this.player) this.player.dispose();
    if (this._sub) this._sub.unsubscribe();
  }
  ...
```

Our audio player can now react to volume changes made via data binding from each track by observing the `volume$` subject observable. Since mute essentially just requires the modification of the player's volume, we ensure that we update the volume accordingly and maintain the original volume if toggling mute on/off, so any custom volume set will be retained.

Our new enriched view of our tracks includes our reusable Waveform view but this time with `type="file"`, since this will engage the audio file static Waveform to be rendered so that we can *see* our audio. We also provide the ability to adjust each track's volume (mixing control) and float a label off to the top-left corner so that the user still knows what is what. This is all done by utilizing an `AbsoluteLayout` container, which allows us to overlap components and manually position them on top of each other.

Serializing data for persistence and hydrating it back upon retrieval

This all works really nicely, however, we have introduced a problem. Our MixerService provides the ability to save these compositions with all their tracks. However the tracks now contain complex objects such as Observables and even private references with getters and setters.

When persisting data, you will often want to use JSON.stringify to serialize objects when storing them so that they can be retrieved later and hydrated into more complex models if needed.

In fact, if you were to attempt to process our TrackModel with JSON.stringify now, it would fail because you cannot stringify certain structures. So, we now need a way to serialize our data before storing it, as well as a way to hydrate that data when retrieving to restore our more sophisticated models. Let's make a few changes to our MixerService to account for this. Open app/modules/mixer/services/mixer.service.tsand make the following changes:

```
// nativescript
import { knownFolders, path } from 'file-system';
...
@Injectable()
export class MixerService {

  public list: Array<IComposition>;

  constructor(
    private databaseService: DatabaseService,
    private dialogService: DialogService
  ) {
    // restore with saved compositions or demo list
    this.list = this._hydrateList(this._savedCompositions() ||
this._demoComposition());
  }
  ...
  private _saveList() {
    this.databaseService.setItem(DatabaseService.KEYS.compositions,
this._serializeList());
  }

  private _serializeList() {
    let serialized = [];
    for (let comp of this.list) {
```

```
      let composition: any = Object.assign({}, comp);
      composition.tracks = [];
      for (let track of comp.tracks) {
        let serializedTrack = {};
        for (let key in track) {
          // ignore observable, private properties and waveform model
(redundant)
          // properties are restored upon hydration
          if (!key.includes('_') && !key.includes('$') && key != 'model') {
            serializedTrack[key] = track[key];
          }
        }
        composition.tracks.push(serializedTrack);
      }
      // serialized composition
      serialized.push(composition);
    }
    return serialized;
  }

  private _hydrateList(list: Array<IComposition>) {
    for (let c = 0; c < list.length; c++) {
      let comp = new CompositionModel(list[c]);
      for (let i = 0; i < comp.tracks.length; i++) {
        comp.tracks[i] = new TrackModel(comp.tracks[i]);
        // for waveform
        (<any>comp.tracks[i]).model = {
          // fix is only for demo tracks since they use files from app
folder
          target: fixAppLocal(comp.tracks[i].filepath)
        };
      }
      // ensure list ref is updated to use hydrated model
      list[c] = comp;
    }
    return list;
  }
  ...
}

const fixAppLocal = function (filepath: string) {
  if (filepath.indexOf('~/') === 0) {
    // needs to be absolute path and not ~/ app local shorthand
    return path.join(knownFolders.currentApp().path, filepath.replace('~/',
''));
  }
  return filepath;
}
```

We will now ensure that any time our composition is saved, it's properly serialized into a safe and more simplified form, which can be processed by `JSON.stringify`. Then, when retrieving data out of our persistent store (in this case, via NativeScript's application-settings module being used under the hood of our `DatabaseService`; this is covered in `Chapter 2`, *Feature Modules*, we hydrate the data back into our models, which will enrich the data with our observable properties.

Leveraging Angular directives to enrich the NativeScript Slider with more unique characteristics

For each track fader (also known as our mixing/volume control), it'd be nice to actually render a fader-looking control knob so that it's clear that these Sliders are mixing knobs and are not mistaken for shuttling playback of that track, for instance. We can create a graphic to be used for these Sliders, which will look like this:

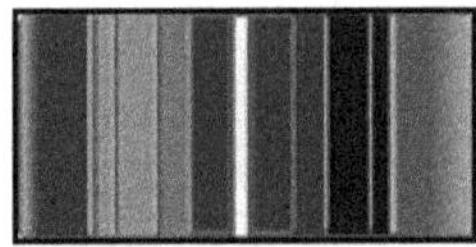

For iOS, we will also want a down/highlighted state, so usability feels good when the user presses down on the fader:

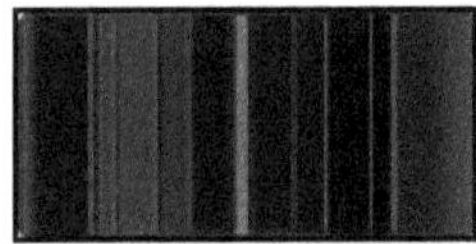

We can now create two versions of each of these files and drop them in `app/App_Resources/iOS`; the original will be 100x48 for standard resolution, then for iPhone Plus and above, we will have a @3x version at 150x72 (basically, 24 plus the standard 48 height):

- `fader-down.png`
- `fader-down@3x.png`
- `fader.png`
- `fader@3x.png`

We can now reuse our `SlimSliderDirective` (currently being used to customize the look of the shuttle slider) and provide an input so that we can provide the name of a file from our app's resources to use for the thumb.
Open `app/modules/player/directives/slider.directive.ios.ts` and make the following modifications:

```
import { Directive, ElementRef, Input } from '@angular/core';

@Directive({
  selector: '[slim-slider]'
})
export class SlimSliderDirective {
  @Input('slim-slider') imageName: string;

  constructor(private el: ElementRef) { }

  ngAfterViewInit() {
    let uiSlider = <UISlider>this.el.nativeElement.ios;
    if (this.imageName) {
      uiSlider.setThumbImageForState(
        UIImage.imageNamed(this.imageName), UIControlState.Normal);
      // assume highlighted state always suffixed with '-down'
      let imgParts = this.imageName.split('.');
      let downImg = `${imgParts[0]}-down.${imgParts[1]}`;
      uiSlider.setThumbImageForState(
        UIImage.imageNamed(downImg), UIControlState.Highlighted);
    } else {
      // used for shuttle control
      uiSlider.userInteractionEnabled = false;
      uiSlider.setThumbImageForState(UIImage.new(), UIControlState.Normal);
    }
  }
}
```

This allows us to specify the filename to be used as the `Slider` thumb on the component itself:

```
<Slider slim-slider="fader.png" minValue="0" maxValue="1"
  width="94%" top="50" left="0"
  [(ngModel)]="track.volume" class="slider fader"></Slider>
```

With this in place, we now have these neat fader controls for iOS when the track mixing view toggle is engaged:

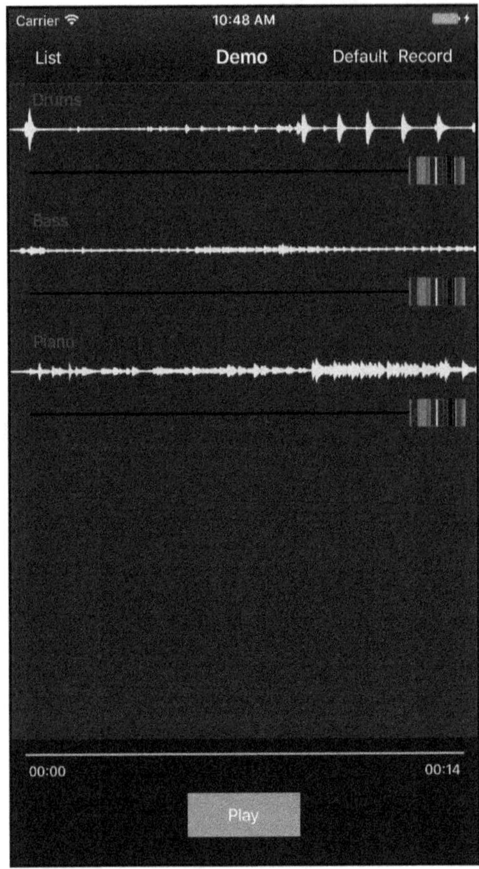

Graphic and resource handling for Android

Now, let's handle this for Android as well. We start by taking our standard 48 height fader graphic and copying it into app/App_Resources/Android/drawable-hdpi folder. We can then create appropriate resolutions of this graphic and copy into the various resolution dependent folders. The one thing to keep in mind with Android is it does **not** use the "@3x" suffix identifiers like iOS does so we just name all of these "fader.png". Here's a view of our graphic in one of the resolution dependent (in this case **hdpi** which handles "high density" screen resolutions) folders:

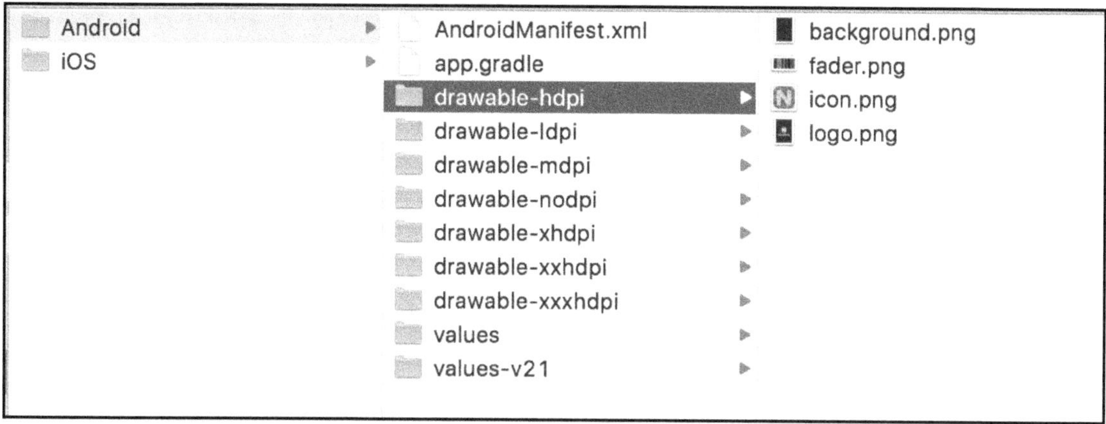

We can now customize our Android slider directive with thumb image handling, open `app/modules/player/directives/slider.directive.android.ts`:

```
import { Directive, ElementRef, Input } from '@angular/core';
import { fromResource } from 'image-source';
import { getNativeApplication } from 'application';

let application: android.app.Application;
let resources: android.content.res.Resources;

const getApplication = function () {
  if (!application) {
    application = (<android.app.Application>getNativeApplication());
  }
  return application;
}

const getResources = function () {
  if (!resources) {
    resources = getApplication().getResources();
  }
  return resources;
}

@Directive({
 selector: '[slim-slider]'
})
export class SlimSliderDirective {
  @Input('slim-slider') imageName: string;
  private _thumb: android.graphics.drawable.BitmapDrawable;

  constructor(private el: ElementRef) {
```

```
    el.nativeElement[(<any>slider).colorProperty.setNative] = function (v)
{
      // ignore the NativeScript default color setter of this slider
    };
  }

  ngAfterViewInit() {
    let seekBar = <android.widget.SeekBar>this.el.nativeElement.android;
    if (this.imageName) {
      if (!seekBar) {
        // part of view toggle - grab on next tick
        // this helps ensure the seekBar instance can be accessed properly
        // since this may fire amidst the view toggle switching on our
tracks
        setTimeout(() => {
          seekBar = <android.widget.SeekBar>this.el.nativeElement.android;
          this._addThumbImg(seekBar);
        });
      } else {
        this._addThumbImg(seekBar);
      }
    } else {
      // seekBar.setEnabled(false);
      seekBar.setOnTouchListener(new android.view.View.OnTouchListener({
        onTouch(view, event) {
          return true;
        }
      }));
      seekBar.getThumb().mutate().setAlpha(0);
    }
  }

  private _addThumbImg(seekBar: android.widget.SeekBar) {
    if (!this._thumb) {
      let imgParts = this.imageName.split('.');
      let name = imgParts[0];
      const res = getResources();
      if (res) {
        const identifier: number = res.getIdentifier(
          name, 'drawable',  getApplication().getPackageName());
        if (0 < identifier) {
          // Load BitmapDrawable with getDrawable to make use of Android
internal caching
          this._thumb =
<android.graphics.drawable.BitmapDrawable>res.getDrawable(identifier);
        }
      }
    }
```

```
      if (this._thumb) {
        seekBar.setThumb(this._thumb);
        seekBar.getThumb().clearColorFilter();
        if (android.os.Build.VERSION.SDK_INT >= 21) {
          (<any>seekBar).setSplitTrack(false);
        }
      }
    }
  }
```

One of the most unusual and perhaps surprising aspects of the Android implementation above is this:

```
constructor(private el: ElementRef) {
  el.nativeElement[(<any>slider).colorProperty.setNative] = function (v) {
    // ignore the NativeScript color setter of the slider
  };
}
```

Normally you can reuse and extend controls in NativeScript quite easily. However, this is one of those exception cases where the default NativeScript setter on the slider control is actually going to cause us problems **only on Android**. The default setter will attempt to set the color of the thumb to blue along with a method to blend it. When it sets this flag on the slider, any graphic shape we then set gets the shape set to all blue. So for our version of the slider class to handle a custom graphic we have to eliminate the default slider thumb color setter on the control. We take control of this by attaching a new "color" setter that does absolutely nothing. This way when the NativeScript framework tries to setup the default color while initializing or resetting the control, nothing will happen allowing us to completely control what happens. As a precaution at the end of the `_addThumbImg` method we also call `seekBar.getThumb().clearColorFilter();` for good measure to make sure any potential sets to the colorFilter are undone before we were able to silence the default color setter.

Lastly we can customize the colors used in the audio waveform shown for each track when our track listing view is toggled to mixer mode. Since the waveform plugin for Android utilizes the color resources of the app, we can add the proper named attributes found in the plugin's documentation in app/App_Resources/Android/values/colors.xml and the same colors should also be copied into app/App_Resources/Android/values-v21/colors.xml:

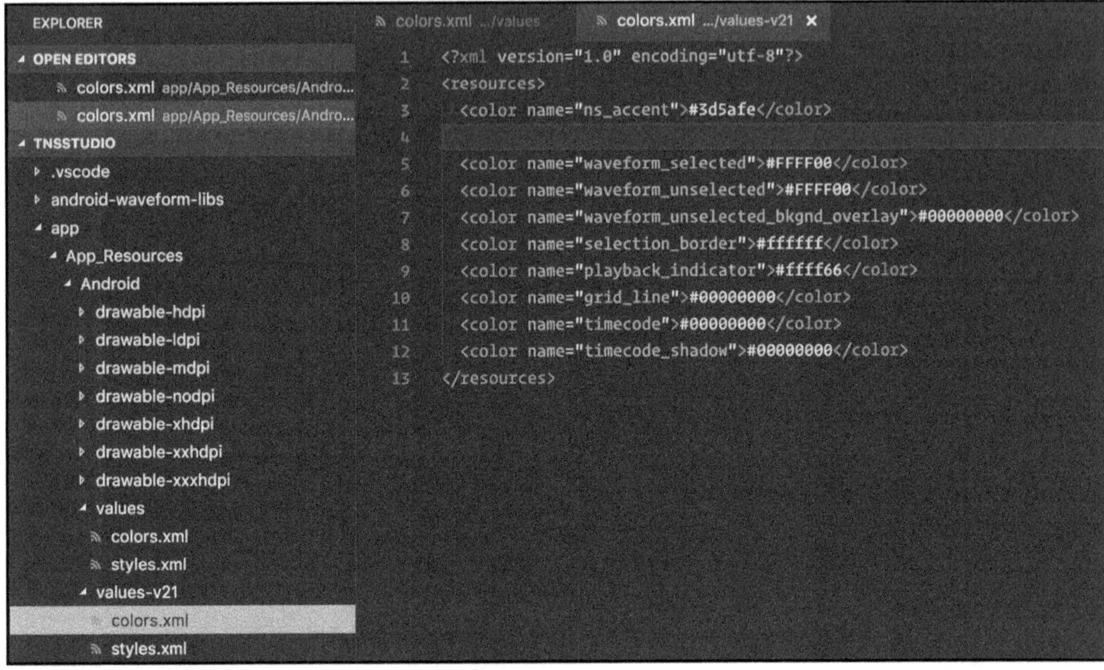

This now gives us a custom style for Android's file waveform display in mixing mode:

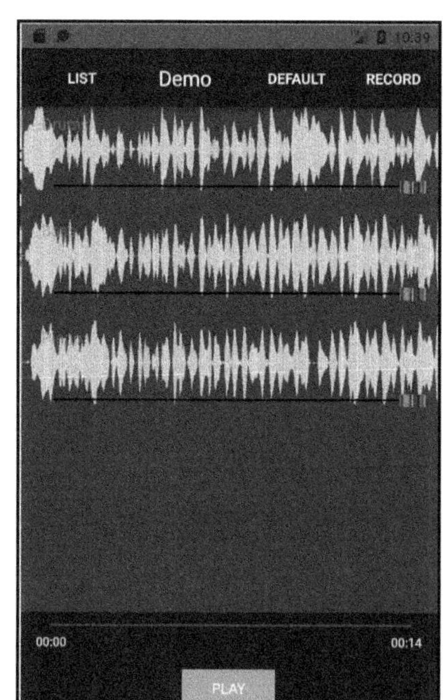

Summary

We wanted to provide some extra goodies to go along with all the rich content you've been learning about throughout Part 3; we hope that you enjoyed it! Using multiple item row templates with `ListView` can be handy in many situations, but hopefully, this will give you the tools to know how to make it work for you and your app.

Special considerations with data persistence is an important factor to any engaging app, so we looked at serializing data before storing and hydrating the data when restoring it out of a persisted state.

Lastly, we looked at further enriching our view components with more Angular directive goodness. With Part 3 completed, we have now completed the core competency and feature set of our app for the book. However, we are far from done with the app. The development workflow and process presented in this book are the typical development cycle we bring to any app we build. We will cover improving our architecture and further polishing our app to prepare for a public release via Google Play and App Store in `Chapter 14`, *Deployment Preparation with webpack Bundling*.

Let's now move on to improving the state handling of our app with `ngrx` integration in `Chapter 10`, *@ngrx/store + @ngrx/effects for State Management*. It's worth mentioning that using a Redux style architecture is a decision better made *before* building out your app as we have done here. However, it's not necessarily critical nor is it mandatory, therefore, we wanted to build the app excluding it to show that the app fundamentally works just fine. Now, we will move on to working it in to show off the various advantages you can gain with it.

10
@ngrx/store + @ngrx/effects for State Management

Managing state in any app can become troubling as the app scales over time. We want to have full confidence over the predictability of our app's behavior and getting a hang of its state is key to gaining that confidence.

State can be broadly defined as the particular condition that someone or something is in at a specific time. With regard to our app, the state can encompass whether our player is playing or not, whether the recorder is recording or not, and whether the track list UI is in mixing mode or not.

Storing state in a single spot allows you to know exactly what the state of the app is at any given moment. Without a single store, you usually wind up with state buried throughout different components and services, which often leads to two or more different versions of state as features are built out. This unwieldy growth of state becomes even more troublesome as different features need to interact with each other, which may or may not necessarily depend on each other.

In this chapter, we will cover the following topics:

- Understanding what Redux is
- Understanding what ngrx is and how it relates to Redux
- Defining state for an app
- Integrating @ngrx/store to manage state
- Understanding what @ngrx/effects are
- Integrating side effects to aid our state management
- Going from *inactive to reactive* with our code base (Mike Ryan/Brandon Roberts™)

Understanding Redux and integrating @ngrx/store

Redux is an open source library that defines itself as a predictable state container for JavaScript apps. The concepts are not exactly new, but the details were developed by Dan Abramov in 2015 who was influenced by Facebook's Flux and the functional programming language, Elm. It quickly gained popularity among the React community as it was used throughout Facebook.

We don't want to redefine what Redux is, so we will quote directly from the Redux repo (`https://github.com/reactjs/redux`):

> The whole state of your app is stored in an object tree inside a single *store*.
> The only way to change the state tree is to emit an *action*, an object describing what happened.
> To specify how the actions transform the state tree, you write pure *reducers*.
>
> That's it!

The concept is fairly simple and quite brilliant. You emit actions (which are simple string typed objects with a payload representing the data to be passed along) against the system, which wind up hitting a reducer (a pure function) to define how state is transformed by these actions.

It's important not to confuse transform with mutate. One of the fundamental concepts of Redux is that all state is **immutable**; hence, each reducer is a **pure** function.

A pure function always returns the same results given the same parameters. Its execution does not depend on the state of the system as a whole [`https://en.wikipedia.org/wiki/Pure_function`].

So, although a reducer transforms state, it does not mutate it.

In depth, engineering studies have been done on change detection systems and how object equality/reference checks are superior in speed when compared to object comparison checks on deeply nested properties. We won't go into detail for the reasons for this, but immutability of your app's data flow has significant impact on how you can fine-tune its performance, especially with regard to Angular.

Along with performance enhancements, the concepts of Redux further enhance decoupling across your entire code base, leading to the reduction of various dependencies spread throughout. With the power of actions describing the various interactions our app entails, we no longer need to inject explicit service dependencies to execute its APIs. Instead, we can simply emit actions and the principles of Redux will work for us to propagate and handle the necessary functionality our app demands, all the while maintaining a single and dependable source of truth.

What is @ngrx/store?

Early in the rewrite of Angular (from 1.x to 2.x+), a core team member turned developer advocate at Google, Rob Wormald, developed **ngrx/store** as an "*RxJS powered state management [system] for Angular applications, inspired by Redux.*" The key point in that phrase is the term "**RxJS**". Hence the name **ngrx** derives its name from joining "**ng**" for A**ng**ular with "**rx**" from **Rx**JS. The open source library quickly gained highly-talented contributors such as Mike Ryan, Brian Troncone, and Brandon Roberts and took off to become an extremely intelligent and powerful state management system for modern Angular applications.

Although it is heavily inspired by Redux and utilizes the same concepts, it is uniquely different in making RxJS a first-class citizen in how the system is wired. It brings **Observables** full circle throughout all the concepts of Redux, enabling truly **reactive** user interfaces and apps.

If all these concepts are new to you, Brian Troncone's thorough post will definitely help you gain more understanding as we won't be able to cover every detail of ngrx here. Please see this post:

- `https://gist.github.com/btroncone/a6e4347326749f938510`

Designing the state model

Before integrating ngrx, it's good to first think about the various aspects of state throughout your app in addition to which module they might pertain to. With our app, here's a reasonable starter list (*not meant to be complete or thorough at this point*):

- CoreModule:
    - user: any; user-related state:
        - recentUsername: string; most recently used successful username
        - current: any; authenticated user (if there is one)
- MixerModule:
    - mixer: any: mixer-related state
        - compositions: Array<IComposition>; list of user-saved compositions
        - activeComposition: CompositionModel; the active composition
- PlayerModule:
    - player: any; various aspects of player state.
        - playing: boolean; whether audio is playing or not.
        - duration: number; total duration of playback.
        - completed: boolean; whether playback reached the end and is completed. This will help determine the difference between when the user stops playback or when it autostops due to the player reaching the end.
        - seeking: boolean; whether playback seeking is in progress.
- RecorderModule:
    - recorder: RecordState; recording state represented simply by an enum

No module in particular, just state we want to observe:

- ui: any; user interface state
    - trackListViewType: string; the currently active view toggle for track listing

The key point here is not to worry about getting this exactly right the first time. It's hard to know the precise state model when you first build an app, and it will most likely change a bit over time and that's okay.

State for our app is better known at this time because we have already built a working app, so this is a tad bit easier. Typically, mapping this out before you build an app is more difficult; however, again, don't worry about getting it right the first time! You can easily refactor and tweak it over time.

Let's take this state and work it into our app with ngrx.

Installing and integrating @ngrx/store

We want to first install `@ngrx/store`:

```
npm i @ngrx/store --save
```

We can now provide the single store to our app via the `StoreModule`. We define these initial slices of state in our `CoreModule`, which will be available when the app boots, while each lazy loaded feature module adds its own state and reducers later when needed.

Providing the initial app state excluding any lazily loaded module state

We want to start by defining the initial app state, excluding any lazily loaded feature module state. Since our `CoreModule` provides `AuthService`, which deals with handling our user, we will consider the **user** slice a fundamental key to our app's initial state.

In particular, let's begin by defining the shape of our user state. Create `app/modules/core/states/user.state.ts`:

```
export interface IUserState {
  recentUsername?: string;
  current?: any;
  loginCanceled?: boolean;
}

export const userInitialState: IUserState = {};
```

Our user state is very simple. It contains a `recentUsername` representing a string of the most recently successfully authenticated username (useful if the user were to log out and return to log in later). Then, we have **current**, which will represent a user object if authenticated, or null if not. We also include a `loginCanceled` boolean since we surmise it may be useful for analyzing user interaction if we were to start reporting state as analytics data.

Any data points around authentication can be critical to understanding our app's user base. For example, it might be insightful to learn whether or not requiring authentication to record was causing more canceled logins than signups, which may have a direct affect on user retention.

To be consistent with our approach throughout this book, also create `app/modules/core/states/index.ts`:

```
export * from './user.state';
```

Now, let's create our user actions; create `app/modules/core/actions/user.action.ts`:

```
import { Action } from '@ngrx/store';
import { IUserState } from '../states';

export namespace

UserActions {
  const CATEGORY: string = 'User';

  export interface IUserActions {
    INIT:

string;
    LOGIN: string;
    LOGIN_SUCCESS: string;
    LOGIN_CANCELED: string;
    LOGOUT:

string;
    UPDATED: string;
  }

  export const ActionTypes: IUserActions = {
    INIT:

`${CATEGORY} Init`,
    LOGIN:          `${CATEGORY} Login`,
    LOGIN_SUCCESS:  `${CATEGORY} Login Success`,
```

```
    LOGIN_CANCELED: `${CATEGORY} Login Canceled`,
    LOGOUT:         `${CATEGORY} Logout`,
    UPDATED:

`${CATEGORY} Updated`
  };

  export class InitAction implements Action {
    type =

ActionTypes.INIT;
    payload = null;
  }

  export class LoginAction implements Action {
    type

= ActionTypes.LOGIN;
    constructor(public payload: { msg: string; usernameAttempt?: string}) {
}
  }

  export class LoginSuccessAction implements Action {
    type = ActionTypes.LOGIN_SUCCESS;
    constructor

(public payload: any /*user object*/) { }
  }

  export class LoginCanceledAction implements Action {

  type = ActionTypes.LOGIN_CANCELED;
    constructor(public payload?: string /*last attempted username*/) { }

}

  export class LogoutAction implements Action {
    type = ActionTypes.LOGOUT;
    payload =

null;
  }

  export class UpdatedAction implements Action {
    type = ActionTypes.UPDATED;

constructor(public payload: IUserState) { }
  }
```

```
    export type Actions =
      InitAction
      |

  LoginAction
      | LoginSuccessAction
      | LoginCanceledAction
      | LogoutAction
      |

  UpdatedAction;
  }
```

Then, follow up with our standard; create `app/modules/core/actions/index.ts`:

```
export * from './user.action';
```

Okay now, what's going on with those actions?! Here's what we've defined:

- `INIT`: To initialize the user right when the app launches. In other words, this action will be used to check persistence and restore a user object onto the app's state at launch time.
- `LOGIN`: Begin the login sequence. In our app, this will show the login dialog.
- `LOGIN_SUCCESS`: Since login is asynchronous, this action will dispatch once login is complete.
- `LOGIN_CANCELED`: If the user cancels login.
- `LOGOUT`: When user logs out.
- `UPDATED`: We will use this as a simple action to update our user state. This will generally not be dispatched directly, but will be used in the reducer we'll create in a moment.

The formalities you see here provide a consistent and strongly-typed structure. By utilizing a namespace, we are able to uniquely identify this set of actions with a name, `UserActions`. This allows the interior naming to remain the same across many other namespaced actions we will create for the lazy loaded modules state, providing a great standard to work with. The `CATEGORY` is necessary because every action must be unique, not just in this set of actions but across the entire app. The interfaces help provide good intelligence when using our actions, in addition to type safety. The various action classes help ensure that all actions dispatched are new instances and provide a powerful way to strongly type our action payloads. This also makes our code easy to refactor down the line. The last utility in our structure is the union type at the bottom, which helps our reducer determine the applicable actions it should be concerned with.

Speaking of that reducer, let's create it now: `app/modules/core/reducers/user.reducer.ts`:

```
import { IUserState, userInitialState } from '../states/user.state';
import { UserActions } from

'../actions/user.action';

export function userReducer(
  state: IUserState = userInitialState,

action: UserActions.Actions
): IUserState {
  switch (action.type) {
    case

UserActions.ActionTypes.UPDATED:
      return Object.assign({}, state, action.payload);
    default:

return state;
  }
}
```

The reducer is incredibly simple. As mentioned, it is a pure function that takes in the existing state along with an action and returns a new state (as a new Object unless it's the default starting case). This maintains immutability and keeps things quite elegant. The `UPDATED` action will always be the last in any action chain to fire off and ultimately change the user state. In this case, we'll keep things simple and allow our `UPDATED` action to be the only action that actually changes the user state. The other actions will set up a chain, whereby they end up dispatching `UPDATED` if they need the user state to change. You could certainly set up more cases here based on our actions to change the state; however, in our app, this will be the only action that ultimately changes the user state.

Action chain? What on earth do we mean by an *Action chain*?! You may be wondering how we wire these actions to interplay with each other if needed?

Installing and integrating @ngrx/effects

Without redefining, let's look at the description of @ngrx/effects straight from the repo (`https://github.com/ngrx/effects`):

In `@ngrx/effects`, effects are the sources of actions. You use the `@Effect()` decorator to hint which observables on a service are action sources, and `@ngrx/effects` automatically merges your action streams, letting you subscribe them to store.

To help you compose new action sources, `@ngrx/effects` exports an action observable service that emits every action dispatched in your application.

In other words, we can chain our actions together with effects to provide powerful data flow composition throughout our app. They allow us to insert behavior that should take place between when an action is dispatched and before the state is ultimately changed. The most common use case is to handle HTTP requests and/or other asynchronous operations; however, they have many useful applications.

To use, let's first install `@ngrx/effects`:

```
npm i @ngrx/effects --save
```

Now let's take a look at what our user actions look like in an effect chain.

Real quickly, though, to remain consistent with our naming structure, let's rename `auth.service.ts` to `user.service.ts`. It helps when we have a naming standard that is consistent across the board.

Now, create `app/modules/core/effects/user.effect.ts`:

```
// angular
import { Injectable } from '@angular/core';

// libs
import { Store, Action } from

'@ngrx/store';
import { Effect, Actions } from '@ngrx/effects';
import { Observable } from

'rxjs/Observable';
```

```
// module
import { LogService } from '../../core/services/log.service';
import {

DatabaseService } from '../services/database.service';
import { UserService } from '../services/user.service';
import { UserActions } from '../actions/user.action';

@Injectable()
export class UserEffects {

  @Effect() init$: Observable<Action> = this.actions$
    .ofType(UserActions.ActionTypes.INIT)
    .startWith(new UserActions.InitAction())
    .map(action => {
      const current =

this.databaseService
        .getItem(DatabaseService.KEYS.currentUser);
      const recentUsername =

this.databaseService
        .getItem(DatabaseService.KEYS.recentUsername);
      this.log.debug(`Current user:

`, current || 'Unauthenticated');
        return new UserActions.UpdatedAction({ current, recentUsername });

});

  @Effect() login$: Observable<Action> = this.actions$
    .ofType

(UserActions.ActionTypes.LOGIN)
    .withLatestFrom(this.store)
    .switchMap(([action, state]) => {

  const current = state.user.current;
      if (current) {
        // user already logged in, just fire

updated
        return Observable.of(
          new UserActions.UpdatedAction({ current })
        );

    } else {
        this._loginPromptMsg = action.payload.msg;
```

```
        const usernameAttempt =

action.payload.usernameAttempt
          || state.user.recentUsername;

        return

Observable.fromPromise(
          this.userService.promptLogin(this._loginPromptMsg,
          usernameAttempt)


        )
        .map(user => (new UserActions.LoginSuccessAction(user)))
        .catch

(usernameAttempt => Observable.of(
          new UserActions.LoginCanceledAction(usernameAttempt)

));
      }
    });

  @Effect() loginSuccess$: Observable<Action> = this.actions$

.ofType(UserActions.ActionTypes.LOGIN_SUCCESS)
    .map((action) => {
      const user = action.payload;


     const recentUsername = user.username;
      this.databaseService
        .setItem

(DatabaseService.KEYS.currentUser, user);
      this.databaseService
        .setItem

(DatabaseService.KEYS.recentUsername, recentUsername);
      this._loginPromptMsg = null; // clear, no longer

needed
      return (new UserActions.UpdatedAction({
        current: user,
        recentUsername,

    loginCanceled: false
      }));
```

```
    });

  @Effect() loginCancel$ = this.actions$

.ofType(UserActions.ActionTypes.LOGIN_CANCELED)
    .map(action => {
      const usernameAttempt =

action.payload;
      if (usernameAttempt) {
        // reinitiate sequence, login failed, retry

return new UserActions.LoginAction({
          msg: this._loginPromptMsg,
          usernameAttempt

});
      } else {
        return new UserActions.UpdatedAction({
          loginCanceled: true

});
      }
    });

  @Effect() logout$: Observable<Action> = this.actions$

.ofType(UserActions.ActionTypes.LOGOUT)
    .map(action => {
      this.databaseService

.removeItem(DatabaseService.KEYS.currentUser);
      return new UserActions.UpdatedAction({
        current:

null
      });
    });

  private _loginPromptMsg: string;

  constructor(
    private

store: Store<any>,
    private actions$: Actions,
    private log: LogService,
    private
```

```
databaseService: DatabaseService,
    private userService: UserService
  ) { }
}
```

We have clarified the intent of our data flow concerning our `UserService` and delegated the responsibility to this effect chain. This allows us to compose our data flow in a clear and consistent manner with a great deal of flexibility and power. For instance, our `InitAction` chain now allows us to automatically initialize the user via the following:

```
.startWith(new UserActions.InitAction())
```

Earlier, we were calling a private method--`this._init()`--inside the service constructor; however, we no longer need explicit calls like that as effects are run and queued up once the module is bootstrapped. The `.startWith` operator will cause the observable to fire off one single time (at the point of module creation), allowing the init sequence to be executed at a particularly opportune time, when our app is initializing. Our initialization sequence is the same as we were previously handling in the service; however, this time we're taking into consideration our new `recentUsername` persisted value (if one exists). We then end the init sequence with a `UserActions.UpdatedAction`:

```
new UserActions.UpdatedAction({ current, recentUsername })
```

Note that there's no effect chain wired to `UserActions.ActionTypes.UPDATED`. This is because there are no side effects that should occur by the time that `Action` occurs. Since there are no more side effects, the observable sequence ends up in the reducer that has a `switch` statement to handle it:

```
export function userReducer(
  state: IUserState = userInitialState,
  action: UserActions.Actions
):

IUserState {
  switch (action.type) {
    case UserActions.ActionTypes.UPDATED:

return Object.assign({}, state, action.payload);
    default:
      return state;
  }
}
```

This takes the payload (which is typed as the shape of the user state, `IUserState`) and overwrites the values in the existing state to return a brand new user state. Importantly, `Object.assign` allows any existing values in the source object to not be overridden unless explicitly defined by the incoming payload. This allows only new incoming payload values to be reflected on our state, while still maintaining the existing values.

The rest of our `UserEffect` chain is fairly self-explanatory. Primarily, it's handling much of what the service was previously handling, with the exception of prompting the login dialog, which the effect chain is utilizing the service method to do. However, it's worth mentioning that we can go so far as to completely remove this service as the contents of the `promptLogin` method can easily be carried out directly in our effect now.

When deciding if you should handle more logic in your effect or a designated service, it really comes down to personal preference and/or scalability. If you have rather lengthy service logic and more than a couple of methods to handle logic while working with effects, creating a designated service will help greatly. You can scale more functionality into the service without diluting the clarity of your effects chain.

Lastly, unit testing will be easier with a designated service with more logic. In this case, our logic is fairly simple; however, we'll leave the `UserService` for example purposes as well as best practice.

Speaking of, let's take a look at how simplified our `UserService` looks now in `app/modules/core/services/user.service.ts`:

```
// angular
import { Injectable } from '@angular/core';

// app
import { DialogService } from

'./dialog.service';

@Injectable()
export class UserService {

  constructor(

private dialogService: DialogService
  ) { }

  public promptLogin(msg: string, username: string = ''):

Promise<any> {
```

```
        return new Promise((resolve, reject) => {
          this.dialogService.login(msg,

    username, '').then((input) => {
            if (input.result) { // result will be false when canceled
              if

    (input.userName && input.userName.indexOf('@') > -1) {
                if (input.password) {

        resolve({
                    username: input.userName,
                    password: input.password

      });
                } else {
                  this.dialogService.alert('You must provide a password.')

        .then(reject.bind(this, input.userName));
                }
              } else {
                // reject,

    passing userName back to try again
                this.dialogService.alert('You must provide a valid email

     address.')
                  .then(reject.bind(this, input.userName));
              }
            } else {

        // user chose cancel
              reject(false);
            }
          });
        });
      }
    }
```

It's much cleaner now. Okay, so how do we let our app know about all this new goodness?

First, let's follow one of our standards by adding an index to our entire core module; add `app/modules/core/index.ts`:

```
export * from './actions';
export * from './effects';
export * from './reducers';
export * from

'./services';
export * from './states';
export * from './core.module';
```

We simply export all the goodies our core module now provides, including the module itself.

Then, open `app/modules/core/core.module.ts` to finish our wiring:

```
// nativescript
import { NativeScriptModule } from 'nativescript-
angular/nativescript.module';
import {

NativeScriptFormsModule } from 'nativescript-angular/forms';
import { NativeScriptHttpModule } from 'nativescript-

angular/http';

// angular
import { NgModule, Optional, SkipSelf } from '@angular/core';

// libs
import { StoreModule } from '@ngrx/store';
import {

EffectsModule } from '@ngrx/effects';

// app
import { UserEffects } from

'./effects';
import { userReducer } from './reducers';
import { PROVIDERS } from

'./services';
import { PROVIDERS as MIXER_PROVIDERS } from '../mixer/services';
import { PROVIDERS as

PLAYER_PROVIDERS } from '../player/services';
```

```
const MODULES: any[] = [
  NativeScriptModule,

NativeScriptFormsModule,
  NativeScriptHttpModule
];

@NgModule({
  imports: [

...MODULES,
    // define core app state
    StoreModule.forRoot({
      user:

userReducer
    }),
    // register core effects

EffectsModule.forRoot([
      UserEffects
    ]),
  ],
  providers: [

...PROVIDERS,
    ...MIXER_PROVIDERS,
    ...PLAYER_PROVIDERS
  ],
  exports: [
    ...MODULES
  ]
})
export class CoreModule {
  constructor (@Optional() @SkipSelf() parentModule: CoreModule)

{
    if (parentModule) {
      throw new Error(
        'CoreModule is already loaded. Import it in the

AppModule only');
    }
  }
}
```

Here we ensure that we define our `user` state key to use the `userReducer` and register it with `StoreModule`. We then call `EffectsModule.forRoot()`, with a collection of singleton effect providers to register like our `UserEffects`.

Now, let's take a look at how this improves the rest of the code base since we were undoubtedly injecting the `UserService` (previously named `AuthService`) in a couple of places.

We were previously injecting `AuthService` in `AppComponent` to ensure that Angular's dependency injection constructed it early on when the app was bootstrapped, creating the necessary singleton our app needed. However, with `UserEffects` automatically running now on bootstrap, which in turn injects (now renamed) `UserService`, we no longer need this rather silly necessity anymore, so, we can update `AppComponent` as follows:

```
@Component({
  moduleId: module.id,
  selector: 'my-app',
  templateUrl: 'app.component.html',
})
export class AppComponent {

  constructor() { // we removed AuthService (UserService) here
```

In one swoop, our code base is now getting smarter and slimmer. Let's keep going to see other benefits of our ngrx integration.

Open `app/auth-guard.service.ts`, and we can now make the following simplifications:

```
import { Injectable } from '@angular/core';
import { Route, CanActivate, CanLoad } from

'@angular/router';

// libs
import { Store } from '@ngrx/store';
import { Subscription } from 'rxjs/Subscription';

// app
import { IUserState,

UserActions } from '../modules/core';

@Injectable()
export class AuthGuard implements

CanActivate, CanLoad {
```

```
  private _sub: Subscription;

  constructor(private

store: Store<any>) { }

  canActivate(): Promise<boolean> {
    return new Promise

((resolve, reject) => {
      this.store.dispatch(
        new

UserActions.LoginAction({ msg: 'Authenticate to record.' })
      );

this._sub = this.store.select(s => s.user).subscribe((state:
      IUserState) => {

  if (state.current) {
          this._reset();
          resolve

(true);
        } else if (state.loginCanceled) {
          this._reset

();
          resolve(false);
        }

});
    });
  }

  canLoad(route: Route): Promise<boolean> {
    // reuse same

logic to activate
    return this.canActivate();
  }

  private _reset() {
    if (this._sub) this._sub.unsubscribe();
  }
}
```

When activating the `/record` route, we dispatch the `LoginAction` every time since we require an authenticated user to use the recording features. Our login effects chain properly handles if the user is already authenticated, so all we need to do is set up our state subscription to react accordingly.

Ngrx is flexible, and how you set up your actions and effects chains is purely up to you.

Providing lazily loaded feature module state

We can now build out the scalable ngrx structure into our various feature modules, which will provide state. Starting with `MixerModule`, let's modify `app/modules/mixer/mixer.module.ts` with the following:

```
...
// libs
import { StoreModule } from '@ngrx/store';
...

@NgModule({
  imports: [
    PlayerModule,
    SharedModule,

NativeScriptRouterModule.forChild(routes),
    StoreModule.forFeature('mixerModule', {

    mixer: {}     // TODO: add reducer when ready
    })
  ],
  ...
})
export class MixerModule { }
```

Here, we are defining what the `MixerModule` state will provide. Now, let's define its shape; create
`app/modules/mixer/states/mixer.state.ts`:

```
import { IComposition } from '../../shared/models';

export interface IMixerState {
  compositions?:

Array<IComposition>;
  activeComposition?: any;
}
```

```
export const mixerInitialState: IMixerState =

{
  compositions: []
};
```

To be consistent with our approach throughout this book, also create `app/modules/mixer/states/index.ts`:

```
export * from './mixer.state';.
```

Now, let's create our mixer actions; create `app/modules/mixer/actions/mixer.action.ts`:

```
import { ViewContainerRef } from '@angular/core';
import { Action } from '@ngrx/store';
import {

IMixerState } from '../states';
import { IComposition, CompositionModel, TrackModel } from
'../../shared/models';

export namespace MixerActions {
  const CATEGORY: string = 'Mixer';

  export interface

IMixerActions {
    INIT: string;
    ADD: string;
    EDIT: string;
    SAVE: string;
    CANCEL:

string;
    SELECT: string;
    OPEN_RECORD: string;
    UPDATE: string;
    UPDATED: string;
  }



  export const ActionTypes: IMixerActions = {
    INIT: `${CATEGORY} Init`,
    ADD: `${CATEGORY}

Add`,
```

```
    EDIT: `${CATEGORY} Edit`,
    SAVE: `${CATEGORY} Save`,
    CANCEL: `${CATEGORY} Cancel`,

SELECT: `${CATEGORY} Select`,
    OPEN_RECORD: `${CATEGORY} Open Record`,
    UPDATE: `${CATEGORY} Update`,


   UPDATED: `${CATEGORY} Updated`,
  };

  export class InitAction implements Action {
    type =

ActionTypes.INIT;
    payload = null;
  }

  export class AddAction implements Action {
    type =

ActionTypes.ADD;
    payload = null;
  }

  export class EditAction implements Action {
    type =

ActionTypes.EDIT;
    constructor(public payload: CompositionModel) { }
  }

  export class SaveAction

implements Action {
    type = ActionTypes.SAVE;
    constructor(public payload?: Array<CompositionModel>)

{ }
  }

  export class CancelAction implements Action {
    type = ActionTypes.CANCEL;

payload = null;
  }
```

```
  export class SelectAction implements Action {
    type = ActionTypes.SELECT;
    constructor(public payload: CompositionModel) { }
  }

  export class OpenRecordAction implements

Action {
    type = ActionTypes.OPEN_RECORD;
    constructor(public payload?: {
      vcRef:

ViewContainerRef, track?: TrackModel
    }) { }
  }

  export class UpdateAction implements Action

{
    type = ActionTypes.UPDATE;
    constructor(public payload: CompositionModel) { }
  }


export class UpdatedAction implements Action {
    type = ActionTypes.UPDATED;
    constructor(public payload:

IMixerState) { }
  }

  export type Actions =
    InitAction
    | AddAction
    |

EditAction
    | SaveAction
    | CancelAction
    | SelectAction
    | OpenRecordAction
    |

UpdateAction
    | UpdatedAction;
}
```

Similar to our UserActions, we will also use an `INIT` action to autoinitialize this state with user-saved compositions (or our sample demo composition to start). Here's a quick rundown:

- `INIT`: To initialize the mixer right when the app launches. Just as we did with `UserActions`, this action will be used to check persistence and restore any user-saved compositions onto the mixer state at launch time.
- `ADD`: Show the add new composition dialog.
- `EDIT`: Edit a composition's name by prompting a dialog.
- `SAVE`: Save compositions.
- `CANCEL`: General action to cancel out of any effect chain.
- `SELECT`: Select a composition. We will use this action to drive the Angular router to the main selected composition view.
- `OPEN_RECORD`: Handle the preparation of opening the recording view, including checking for authentication, pausing playback if playing, and opening in modal or routing to it.
- `UPDATE`: Initiate an update to a composition.
- `UPDATED`: This will generally not be dispatched directly, but used at the end of an effect sequence that the reducer will pick up to finally change the mixer state.

Now, we can create the reducer that is similar to our user reducer:

```
import { IMixerState, mixerInitialState } from '../states';
import { MixerActions } from '../actions';

export function mixerReducer(
  state: IMixerState = mixerInitialState,
  action: MixerActions.Actions
):

IMixerState {
  switch (action.type) {
    case MixerActions.ActionTypes.UPDATED:
      return

Object.assign({}, state, action.payload);
    default:
      return state;
  }
}
```

After this, let's create our `MixerEffects` at `app/modules/mixer/effects/mixer.effect.ts`:

```
// angular
import { Injectable, ViewContainerRef } from '@angular/core';

// nativescript
import { RouterExtensions } from 'nativescript-angular/router';

// libs
import { Store, Action } from

'@ngrx/store';
import { Effect, Actions } from '@ngrx/effects';
import { Observable } from

'rxjs/Observable';

// module
import { CompositionModel } from '../../shared/models';
import {

PlayerActions } from '../../player/actions';
import { RecordComponent } from

'../../recorder/components/record.component';
import { MixerService } from '../services/mixer.service';
import {

MixerActions } from '../actions';

@Injectable()
export class MixerEffects {

  @Effect()

init$: Observable<Action> = this.actions$
    .ofType(MixerActions.ActionTypes.INIT)
    .startWith(new

MixerActions.InitAction())
    .map(action =>
      new MixerActions.UpdatedAction({

compositions: this.mixerService.hydrate(
        this.mixerService.savedCompositions()
        ||
```

```
this.mixerService.demoComposition())
      })
    );

  @Effect() add$: Observable<Action> =

this.actions$
    .ofType(MixerActions.ActionTypes.ADD)
    .withLatestFrom(this.store)
    .switchMap

(([action, state]) =>
      Observable.fromPromise(this.mixerService.add())
        .map(value => {

        if (value.result) {
             let compositions = [...state.mixerModule.mixer.compositions];

let composition = new CompositionModel({
              id: compositions.length + 1,
              name:

value.text,
              order: compositions.length // next one in line
            });

compositions.push(composition);
            // persist changes
            return new MixerActions.SaveAction

(compositions);
          } else {
            return new MixerActions.CancelAction();
          }

   })
    );

  @Effect() edit$: Observable<Action> = this.actions$
    .ofType

(MixerActions.ActionTypes.EDIT)
    .withLatestFrom(this.store)
    .switchMap(([action, state]) => {

  const composition = action.payload;
      return
Observable.fromPromise(this.mixerService.edit(composition.name))
```

```
        .map(value => {
          if (value.result) {
            let compositions =

[...state.mixerModule.mixer.compositions];
            for (let i = 0; i < compositions.length; i++) {

      if (compositions[i].id === composition.id) {
                compositions[i].name = value.text;

    break;
              }
            }
            // persist changes
            return new

MixerActions.SaveAction(compositions);
          } else {
            return new MixerActions.CancelAction();
          }
        })
      });

  @Effect() update$: Observable<Action> = this.actions

$
    .ofType(MixerActions.ActionTypes.UPDATE)
    .withLatestFrom(this.store)
    .map(([action, state])

=> {
      let compositions = [...state.mixerModule.mixer.compositions];
      const composition =

action.payload;
      for (let i = 0; i < compositions.length; i++) {
        if (compositions[i].id ===

composition.id) {
          compositions[i] = composition;
          break;
        }
      }

 // persist changes
      return new MixerActions.SaveAction(compositions);
    });
```

```
  @Effect()

select$: Observable<Action> = this.actions$
    .ofType(MixerActions.ActionTypes.SELECT)
    .map(action

=> {
      this.router.navigate(['/mixer', action.payload.id]);
      return new MixerActions.UpdatedAction

({
        activeComposition: action.payload
      });
    });

  @Effect({ dispatch: false })

openRecord$: Observable<Action> = this.actions$
    .ofType(MixerActions.ActionTypes.OPEN_RECORD)

.withLatestFrom(this.store)
    // always pause/reset playback before handling
    .do(action => new

PlayerActions.PauseAction(0))
    .map(([action, state]) => {
      if

(state.mixerModule.mixer.activeComposition &&
        state.mixerModule.mixer.activeComposition.tracks.length)

{
        // show record modal but check authentication
        if (state.user.current) {
          if

(action.payload.track) {
            // rerecording
            this.dialogService
              .confirm

(
                'Are you sure you want to re-record this track?'
              ).then((ok) => {

        if (ok)
                this._showRecordModal(
                  action.payload.vcRef,
```

```
              action.payload.track
                    );
                  });
                } else {

  this._showRecordModal(action.payload.vcRef);
                }
              } else {
                this.store.dispatch(
                  new UserActions.LoginToRecordAction(action.payload));
              }
            } else {
              //

  navigate to it
              this.router.navigate(['/record']);
            }
            return action;
          });

    @Effect() save$: Observable<Action> = this.actions$
      .ofType(MixerActions.ActionTypes.SAVE)

  .withLatestFrom(this.store)
      .map(([action, state]) => {
        const compositions = action.payload ||
                             state.mixerModule.mixer.compositions;
        // persist
        this.mixerService.save

  (compositions);
        return new MixerActions.UpdatedAction({ compositions });
      });

    constructor

  (
      private store: Store<any>,
      private actions$: Actions,
      private router:

  RouterExtensions,
      private dialogService: DialogService,
      private mixerService: MixerService
    ) { }
```

```
  private _showRecordModal(vcRef: ViewContainerRef, track?: TrackModel) {
    let context: any = {

isModal: true };
    if (track) {
      // re-recording track
      context.track = track;
    }

  this.dialogService.openModal(
      RecordComponent,
      vcRef,
      context,

'./modules/recorder/recorder.module#RecorderModule'
    );
  }
}
```

Probably, the most interesting effect is the `openRecord$` chain. We use `@Effect({ dispatch: false })` to indicate that it should not dispatch any actions at the end as we are using it to execute work directly, such as checking whether the user is authenticated or if `activeComposition` contains tracks to conditionally open record view in a modal or as a route. We make use of another operator:

```
.do(action => new PlayerActions.PauseAction(0))
```

This allows us to insert an arbitrary action without affecting the sequence of events. In this case, we ensure that playback is always paused when the user attempts to open a record view (since they can attempt to open the record view while playback is playing). We are presenting a few more advanced usage options with this chain, just to show what is possible. We are also stepping ahead a bit since we have not shown the creation of `PlayerActions` yet; however, we will just be presenting a couple of highlights in this chapter.

With this effect chain, we can simplify our `MixerService` with the following:

```
...
@Injectable()
export class MixerService {
  ...
  public add() {
    return

this.dialogService.prompt('Composition name:');
  }
```

```
  public edit(name: string) {

return this.dialogService.prompt('Edit name:', name);
  }
  ...
```

We've simplified the service logic, leaving most of the result handling work inside the effects chain. You might decide to leave the service with more logic and keep the effects chain simpler; however, we have designed this setup as an example to show more alternate setups with how flexible ngrx is.

To finish up our lazy loaded state handling, ensure that these effects are run; when `MixerModule` loads, we can make these adjustments to the module:

```
...
// libs
import { StoreModule } from '@ngrx/store';
import { EffectsModule } from

'@ngrx/effects';
...
import { MixerEffects } from './effects';
import

{ mixerReducer } from './reducers';

@NgModule({
  imports: [
    PlayerModule,

SharedModule,
    NativeScriptRouterModule.forChild(routes),
    // mixer state
    StoreModule.forFeature

('mixerModule', {
      mixer: mixerReducer
    }),
    // mixer effects

EffectsModule.forFeature([
      MixerEffects
    ])
  ],
  ...
})
export
```

```
class MixerModule { }
```

Now, let's look at how this improves our component handling, starting with `app/modules/mixer/components/mixer.component.ts`:

```
...
export class MixerComponent implements OnInit, OnDestroy {
  ...
  constructor(

private store: Store<any>,
    private vcRef: ViewContainerRef
  ) { }

  ngOnInit()

{
    this._sub = this.store.select(s => s.mixerModule.mixer)
      .subscribe

((state: IMixerState) => {
        this.composition = state.activeComposition;
      });
  }

  public record(track?: TrackModel) {

this.store.dispatch(new MixerActions.OpenRecordAction({
      vcRef: this.vcRef,
      track
    }));
  }

  ngOnDestroy() {
    this._sub.unsubscribe();
  }
}
```

This time, inside `ngOnInit`, we just set up the component to be reactive to our mixer's state by setting the composition to the `activeComposition`. This is guaranteed to always be whichever composition the user has currently selected and is working on. We dispatch our `OpenRecordAction` inside the `record` method, passing along the proper `ViewContainerRef` and a track if the user is rerecording.

Next up is the simplification of `app/modules/mixer/components/mix-list.component.ts`:

```
// angular
import { Component } from '@angular/core';

// libs
import { Store } from

'@ngrx/store';
import { Observable } from 'rxjs/Observable';

// app
import { MixerActions } from '../actions';
import { IMixerState } from '../states';

@Component({
  moduleId: module.id,
  selector: 'mix-list',
  templateUrl: 'mix-list.component.html'
})
export class MixListComponent {
  public mixer$: Observable<IMixerState>;

  constructor(private store: Store<any>) {
    this.mixer$ = store.select(s => s.mixerModule.mixer);
  }

  public add() {
    this.store.dispatch(new MixerActions.AddAction());
  }

  public edit(composition) {
    this.store.dispatch(new MixerActions.EditAction(composition));
  }

  public select(composition) {
    this.store.dispatch(new MixerActions.SelectAction(composition));
  }
}
```

We have removed the `MixerService` injection and made it reactive by setting up a state Observable--`mixer$`--and integrated our `MixerActions`. This lightens up the component, making it easier to test and maintain since it no longer has an explicit dependency on the `MixerService`, which was previously being used for view bindings as well. If we take a look at the view, we can now utilize Angular's async pipe to gain access to the user-saved compositions via the state:

```
<ActionBar title="Compositions" class="action-bar">
  <ActionItem (tap)="add()"

ios.position="right">
    <Button text="New" class="action-item"></Button>

</ActionItem>
</ActionBar>
<ListView [items]="(mixer$ | async)?.compositions |

orderBy: 'order'" class="list-group">
  <ng-template let-composition="item">
    <GridLayout

rows="auto" columns="100,*,auto" class="list-group-item">
      <Button text="Edit" (tap)="edit(composition)"

row="0" col="0"></Button>
      <Label [text]="composition.name" (tap)="select(composition)" row="0"

col="1" class="h2"></Label>
      <Label [text]="composition.tracks.length" row="0" col="2"
class="text-

right"></Label>
    </GridLayout>
  </ng-template>
</ListView>
```

From the official documentation: Angular's async pipe subscribes to an Observable or Promise and returns the latest value it has emitted. When a new value is emitted, the async pipe marks the component to be checked for changes. When the component gets destroyed, the async pipe unsubscribes automatically to avoid potential memory leaks.

This is truly remarkable and incredibly handy, allowing us to create reactive components that're highly maintainable and flexible.

Inspect the code! Exploring more on your own

Since a lot of what we saw earlier are the exact same principles applied to the rest of our code base, instead of increasing the size of this book further, we invite you to explore the rest of the ngrx integration in the same chapter branch on the accompanying code repository to this book.

Looking through the actual code, running it, and even stepping through it will hopefully give you a solid understanding of how ngrx fits into your app and the many advantages it can bring.

The community is lucky to have members like Rob Wormald, Mike Ryan, Brian Troncone, Brandon Roberts, and more, who have helped make ngrx so nice to use, so a **huge thank you to all the contributors**!

Summary

Hopefully, you are starting to see a pattern of simplification and clarity to the data flow while integrating ngrx. It has helped reduce code, while improving data flow by providing consistent effect chains to various actions, which may need to occur anywhere (from lazy loaded modules or not). By reducing the overhead of managing explicit injected dependencies throughout and instead relying on Store and Actions to initiate the appropriate work, we are increasing the maintainability and scalability of our app. On top of all that, it is paving a pathway to effective testability, which we will cover in `Chapter 12`, *Unit Testing*.

This chapter highlighted the additional advantages when combining NativeScript with Angular by opening up integration potential with rich libraries such as ngrx to improve our app's architecture and data flow.

It's been a long time coming, and we couldn't be more excited about `Chapter 11`, *Polish with SASS*, coming up next. Finally, we are ready to polish our app to give it that special spark!

11
Polish with SASS

After covering some under the hood plumbing improvements with ngrx state management in the previous chapter, it's finally time to polish this app to improve it's overall look and feel. The timing of styling is completely up to your flow of development, and oftentimes, we like to polish as we go. In this book, we chose to avoid intermixing polishing via CSS with feature development to keep the concepts more focused. However, now that we're here, we couldn't be more excited about getting that nice look on our app.

Since standard CSS can become burdensome to maintain as styling grows, we will integrate SASS for help. In fact, we will utilize a community plugin developed by the man who helped come up with the NativeScript brand name itself, Todd Anglin.

In this chapter, we will cover the following topics:

- Integrating SASS into your app
- Best practices when building the core theme's SASS setup
- Building a scalable styling setup to maximize style reuse across iOS and Android
- Using font icons, such as *Font Awesome*, using the nativescript-ngx-fonticon plugin

Polishing with SASS

> *SASS is the most mature, stable, and powerful professional grade CSS extension language in the world... Sass is an extension of CSS that adds power and elegance to the basic language. It allows you to use variables, nested rules, mixins, inline imports, and more, all with a fully CSS-compatible syntax. SASS helps keep large stylesheets well-organized and get small stylesheets up and running.*

- http://sass-lang.com/documentation/file.SASS_REFERENCE.html

Sounds good? You bet.

We will first want to install a community plugin published by Todd Anglin:

```
npm install nativescript-dev-sass --save-dev
```

This plugin will set up a hook that will autocompile your SASS to CSS before building your app, so you don't need to worry about installing any other build tools.

We now want to organize our SASS source files in a particular way that will lend itself to ease of maintenance for not only shared styles between iOS and Android, but also easily allow platform-specific tweaks/overrides. The core theme installed by default (`nativescript-theme-core`) ships with a complete set of SASS source files, which are already organized to help you build your custom SASS on top of it easily.

Let's start by renaming the following:

- `app.ios.css` to `app.ios.`**`scss`**
- `app.android.css` to `app.android.`**`scss`**

Then for the contents of `app.ios.scss`:

```
@import 'style/common';
@import 'style/ios-overrides';
```

And for `app.android.scss`:

```
@import 'style/common';
@import 'style/android-overrides';
```

Okay, now, let's create that `style` folder with the various partial SASS import files to aid our setup, starting with the variables:

- `style/_variables.scss`:

  ```
  // baseline theme colors
  @import '~nativescript-theme-core/scss/dark';
  // define our own variables or simply override those from the light
  set here...
  ```

  There are actually many different skins/colors you could base your app's style sheets on. Check out the following section of the docs to see what's available: `http://docs.nativescript.org/ui/theme#color-schemes`. For our app, we will base our colors off the *dark* skin.

Now, create the common shared SASS file, which is where the bulk of the shared styles will go. In fact, we will take everything we had defined in the `common.css` file and place them here (thereafter, removing the `common.css` file we had before):

- `style/_common.scss`:

```
// customized variables
@import 'variables';
// theme standard rulesets
@import '~nativescript-theme-core/scss/index';
// all the styles we had created previously in common.css migrated
into here:

.action-bar {
  background-color:#101B2E; // we can now convert this to a SASS
variable
}

Page {
  background-color:#101B2E; // we can now convert this to a SASS
variable
}

ListView {
  separator-color: transparent;
}

.track-name-float {
  color: RGBA(136, 135, 3, .5); // we can now convert this to a
SASS variable
}

.slider.fader {
  background-color: #000; // we could actually use $black from core
theme now
}

.list-group .muted {
  opacity:.2;
}
```

This uses our variables file we just created, which enables us to provide our own baseline variables from the core theme with our own custom tweaks to the color.

Now, create the Android override file in case we need it:

- `styles/_android-overrides.scss`:

```
@import '~nativescript-theme-core/scss/platforms/index.android';
// our custom Android overrides can go here if needed...
```

This imports the Android overrides from the core theme while still allowing us to apply our own custom overrides if needed.

We can now do the same for iOS:

- `styles/_ios-overrides.scss`:

```
@import '~nativescript-theme-core/scss/platforms/index.ios';
// our custom iOS overrides can go here if needed...
```

Lastly, we can now convert any component-specific `.css` files to **`.scss`**. We had one component using its own defined styles, `record.component.css`. Just rename it to **`.scss`**. The NativeScript SASS plugin will autocompile any nested `.scss` files it finds.

There are two more things you may want to do:
Ignore all `*.css` files from git in addition to hiding `.css` and `.js` files in your IDE.
You don't want to end up with merge conflicts in the future with other developers since your `.css` files will all be generated fresh via the SASS compilation each time you build the app.

Add the following to your `.gitignore` file:

```
*.js
*.map
*.css
hooks
lib
node_modules
/platforms
```

Then, to hide `.js` and `.css` files in VS Code, we could do this:

```
{
  "files.exclude": {
    "**/app/**/*.css": {
      "when": "$(basename).scss"
    },
    "**/app/**/*.js": {
```

```
            "when": "$(basename).ts"
        },
        "**/hooks": true,
        "**/node_modules": true,
        "platforms": true
    }
}
```

Here's a screenshot of what the structure should look like now:

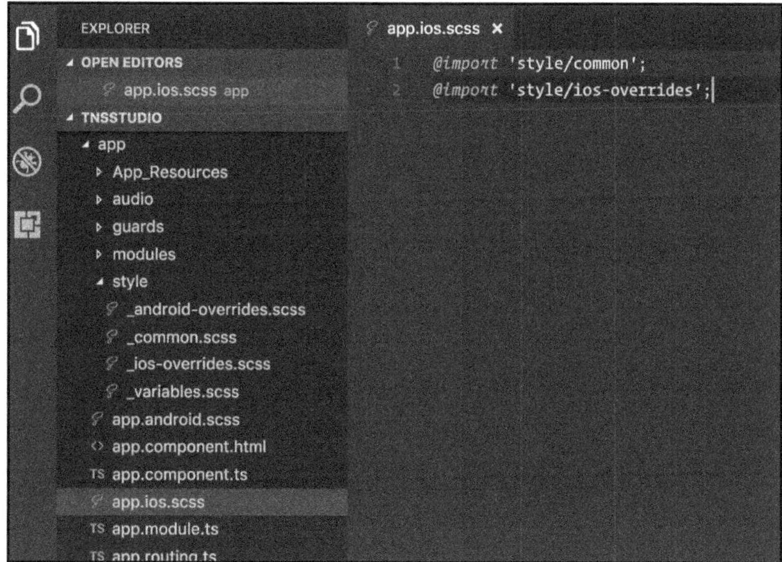

Using font icons with the nativescript-ngx-fonticon plugin

It sure would be nice to swap all those boring label buttons out with nice crisp icons, so let's do that. NativeScript provides support for custom font icons out of the box using Unicode values on text attributes on buttons, labels, and much more. However, with Angular, we can take advantage of another nifty plugin, which will provide a nice Pipe allowing us to use the font names for ease of use and clarity.

Install the following plugin:

```
npm install nativescript-ngx-fonticon --save
```

For this app, we will use the versatile font-awesome icons, so let's download that package here from the official site: `http://fontawesome.io/`.

Inside of it, we will find the font files and the css we will need. We want to first copy the `fontawesome-webfont.ttf` file into a `new fonts` folder we will create in the `app` folder. NativeScript will look for any custom font files in that folder when it builds the app:

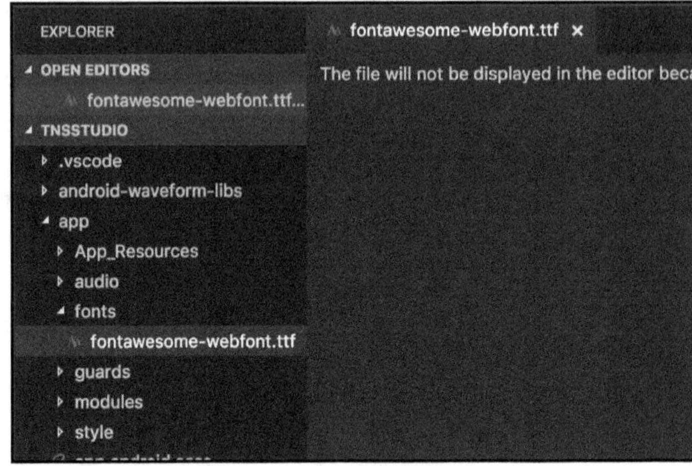

We now want to copy the `css/font-awesome.css` file into our app folder as well. We can either place it in the root of the folder or in a subfolder. We will create an `assets` folder to house this and potentially other similar items in the future.

However, we need to modify this `.css` file slightly. The `nativescript-ngx-fonticon` plugin only works with the font class names and doesn't need any of the utility classes font-awesome provides. So, we will need to modify it to remove a lot of what was at the top to look like this instead:

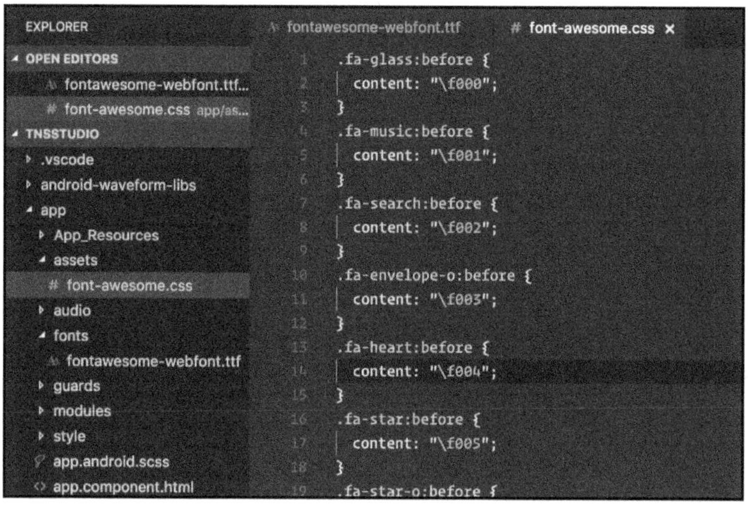

You can learn more about this in the following video: `https://www.youtube.com/watch?v=qb2sk0XXQDw`.

We also set up git to ignore all `*.css` files previously; however, we don't want to ignore the following file:

```
*.js
*.map
*.css
!app/assets/font-awesome.css
hooks
lib
node_modules
/platforms
```

Now, we are ready to set up the plugin. Since this should be part of the core setup of our app, we will modify `app/modules/core/core.module` with our plugin configuration:

```
...
import { TNSFontIconModule } from 'nativescript-ngx-fonticon';
...
@NgModule({
  imports: [
    ...MODULES,
    // font icons
    TNSFontIconModule.forRoot({
      'fa': './assets/font-awesome.css'
    }),
    ...
```

```
  ],
  ...
})
export class CoreModule {
```

Since the module relies on the `TNSFontIconService`, let's modify our root component to inject it, making sure that Angular's DI instantiates the singleton for us to use app-wide.

`app/app.component.ts`:

```
...
// libs
import { TNSFontIconService } from 'nativescript-ngx-fonticon';

@Component({
  moduleId: module.id,
  selector: 'my-app',
  templateUrl: 'app.component.html'
})
export class AppComponent {

  constructor(private fontIconService: TNSFontIconService) {
    ...
```

Next, we want to make sure that the `fonticon` pipe is accessible to any of our view components, so let's import and export the module from our `SharedModule` at `app/modules/shared/shared.module.ts`:

```
...
// libs
import { TNSFontIconModule } from 'nativescript-ngx-fonticon';
...
@NgModule({
  imports: [
    NativeScriptModule,
    NativeScriptRouterModule,
    NativeScriptFormsModule,
    TNSFontIconModule
  ],
  ...
  exports: [
    ...
    TNSFontIconModule,
    ...PIPES
  ]
})
export class SharedModule {}
```

Lastly, we will need a class to designate which components should be used from font-awesome to render themselves. Since this class will be shared across iOS/Android, modify `app/style/_common.scss` with the following:

```
// customized variables
@import 'variables';
// theme standard rulesets
@import '~nativescript-theme-core/scss/index';

.fa {
  font-family: 'FontAwesome', fontawesome-webfont;
  font-size: 25;
}
```

The reason we define two font families is because of the differences between iOS and Android. Android uses the name of the file for the font-family (in this case, `fontawesome-webfont.ttf`). Whereas iOS uses the actual name of the font; examples can be found at `https://github.com/FortAwesome/Font-Awesome/blob/master/css/font-awesome.css#L8`. You *could* rename the font filename to `FontAwesome.ttf` to use just `font-family: FontAwesome` if you wanted. You can learn more at `http://fluentreports.com/blog/?p=176`.

Now, let's try out this new ability to render icons in our app. Open `app/modules/mixer/components/mix-list.component.html`:

```
<ActionBar title="Compositions" class="action-bar">
  <ActionItem (tap)="add()" ios.position="right">
    <Button [text]="'fa-plus' | fonticon" class="fa action-item"></Button>
  </ActionItem>
</ActionBar>
<ListView [items]="(mixer$ | async)?.compositions | orderBy: 'order'"
class="list-group">
  <ng-template let-composition="item">
    <GridLayout rows="auto" columns="100,*,auto" class="list-group-item">
      <Button [text]="'fa-pencil' | fonticon" (tap)="edit(composition)"
        row="0" col="0" class="fa"></Button>
      <Label [text]="composition.name" (tap)="select(composition)"
        row="0" col="1" class="h2"></Label>
      <Label [text]="composition.tracks.length"
        row="0" col="2" class="text-right"> </Label>
    </GridLayout>
  </ng-template>
</ListView>
```

Let's also tweak the background color of our `ListView` to be black for now. We can even use predefined variables from the core theme now with SASS in `app/style/_common.scss`:

```
.list-group {
  background-color: $black;
  .muted {
    opacity:.2;
  }
}
```

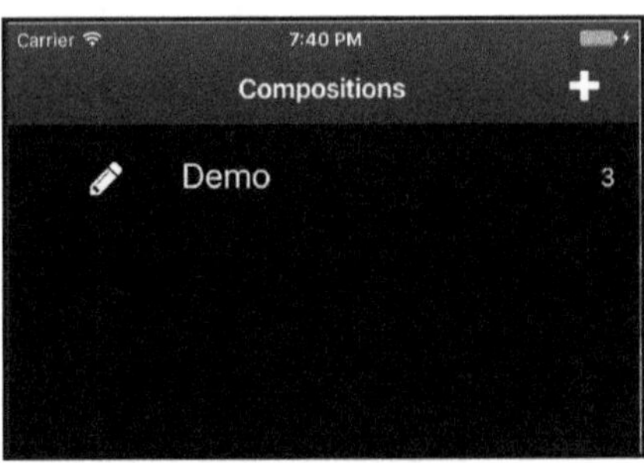

Our composition listing view is now starting to look pretty decent.

Let's keep going and add some icons to our track listing view in `app/modules/player/components/track-list/track-list.component.html`:

```
<ListView #listview [items]="tracks | orderBy: 'order'"
  class="list-group" [itemTemplateSelector]="templateSelector">
  <ng-template let-track="item" nsTemplateKey="default">
    <GridLayout rows="auto" columns="60,*,30"
      class="list-group-item" [class.muted]="track.mute">
      <Button [text]="'fa-circle' | fonticon"
        (tap)="record(track)" row="0" col="0" class="fa c-ruby"></Button>
      <Label [text]="track.name" row="0" col="1" class="h2"></Label>
      <Label [text]="(track.mute ? 'fa-volume-off' : 'fa-volume-up') |
fonticon"
        row="0" col="2" class="fa" (tap)="track.mute=!track.mute"></Label>
    </GridLayout>
  </ng-template>
  ...
```

We are swapping out the Switch we had with a Label designed to toggle two different icons instead. We are also taking advantage of the core theme's handy color classes like c-ruby.

We can also improve our custom `ActionBar` template with some icons:

```
<ActionBar [title]="title" class="action-bar">
  <ActionItem nsRouterLink="/mixer/home">
    <Button [text]="'fa-list-ul' | fonticon" class="fa action-
item"></Button>
  </ActionItem>
  <ActionItem (tap)="toggleList()" ios.position="right">
    <Button [text]="((uiState$ | async)?.trackListViewType == 'default' ?
'fa-sliders' : 'fa-list') | fonticon" class="fa action-item"></Button>
  </ActionItem>
  <ActionItem (tap)="recordAction.next()" ios.position="right">
    <Button [text]="'fa-circle' | fonticon" class="fa c-ruby action-
item"></Button>
  </ActionItem>
</ActionBar>
```

We can now style up the player controls at `app/modules/player/components/player-controls/player-controls.component.html`:

```
<StackLayout row="1" col="0" class="controls">
  <shuttle-slider></shuttle-slider>
  <Button [text]="((playerState$ | async)?.player?.playing ? 'fa-pause' :
'fa-play') | fonticon" (tap)="togglePlay()" class="fa c-white
t-30"></Button>
</StackLayout>
```

We will take advantage of more helper classes from the core theme. The `c-white` class turns our icon white, and `t-30` sets the `font-size: 30`. The latter is short for `text-30`, and the other `color-white`.

Let's take a look:

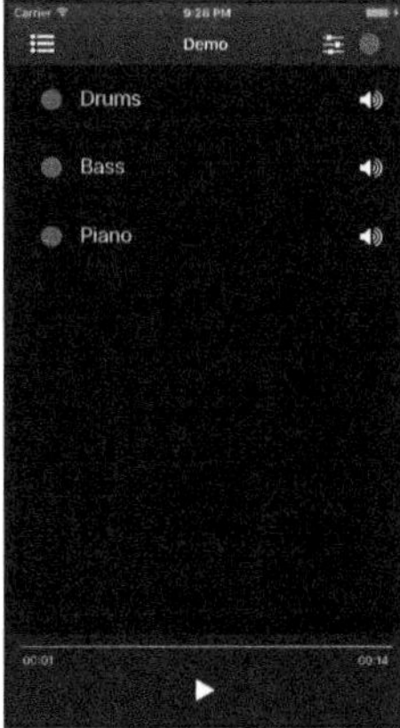

It's amazing how some styling polish can really bring out the personality of your app. Let's crack out the brush one more time on our record view at `app/modules/recorder/components/record.component.html`:

```
<ActionBar title="Record" icon="" class="action-bar">
  <NavigationButton visibility="collapsed"></NavigationButton>
  <ActionItem text="Cancel" ios.systemIcon="1"
(tap)="cancel()"></ActionItem>
</ActionBar>
<FlexboxLayout class="record">
  <GridLayout rows="auto" columns="auto,*,auto" class="p-10"
[visibility]="isModal ? 'visible' : 'collapsed'">
    <Button [text]="'fa-times' | fonticon" (tap)="cancel()" row="0" col="0"
class="fa c-white"></Button>
  </GridLayout>
  <Waveform class="waveform"
    [model]="recorderService.model"
    type="mic"
    plotColor="yellow"
    fill="false"
    mirror="true"
    plotType="buffer">
  </Waveform>
  <StackLayout class="p-5">
    <FlexboxLayout class="controls">
      <Button [text]="'fa-backward' | fonticon" class="fa text-center"
(tap)="recorderService.rewind()" [isEnabled]="state ==
recordState.readyToPlay || state == recordState.playing"></Button>
      <Button [text]="recordBtn | fonticon" class="fa record-btn text-
center" (tap)="recorderService.toggleRecord()" [isEnabled]="state !=
recordState.playing" [class.is-recording]="state ==
recordState.recording"></Button>
      <Button [text]="playBtn | fonticon" class="fa text-center"
(tap)="recorderService.togglePlay()" [isEnabled]="state ==
recordState.readyToPlay || state == recordState.playing"></Button>
    </FlexboxLayout>
    <FlexboxLayout class="controls bottom" [class.recording]="state ==
recordState.recording">
      <Button [text]="'fa-check' | fonticon" class="fa" [class.save-
ready]="state == recordState.readyToPlay" [isEnabled]="state ==
recordState.readyToPlay" (tap)="recorderService.save()"></Button>
    </FlexboxLayout>
  </StackLayout>
</FlexboxLayout>
```

We can adjust our component class to handle the `recordBtn` and `playBtn` now:

```
...
export class RecordComponent implements OnInit, OnDestroy {
  ...
  public recordBtn: string = 'fa-circle';
  public playBtn: string = 'fa-play';
```

Then, to paint everything into place, we can add this to our `app/modules/recorder/components/record.component.scss`:

```
@import '../../../style/variables';

.record {
  background-color: $slate;
  flex-direction: column;
  justify-content: space-around;
  align-items: stretch;
  align-content: center;
}

.record .waveform {
  background-color: transparent;
  order: 1;
  flex-grow: 1;
}

.controls {
  width: 100%;
  height: 200;
  flex-direction: row;
  flex-wrap: nowrap;
  justify-content: center;
  align-items: center;
  align-content: center;

  .fa {
    font-size: 40;
    color: $white;

    &.record-btn {
      font-size: 70;
      color: $ruby;
      margin: 0 50 0 50;

      &.is-recording {
        color: $white;
      }
```

```
      }
    }
  }

  .controls.bottom {
    height: 90;
    justify-content: flex-end;
  }

  .controls.bottom.recording {
    background-color: #B0342D;
  }

  .controls.bottom .fa {
    border-radius: 60;
    font-size: 30;
    height: 62;
    width: 62;
    padding: 2;
    margin: 0 10 0 0;
  }

  .controls.bottom .fa.save-ready {
    background-color: #42B03D;
  }

  .controls .btn {
    color: #fff;
  }

  .controls .btn[isEnabled=false] {
    background-color: transparent;
    color: #777;
  }
```

With this polish, we now have the following screenshot:

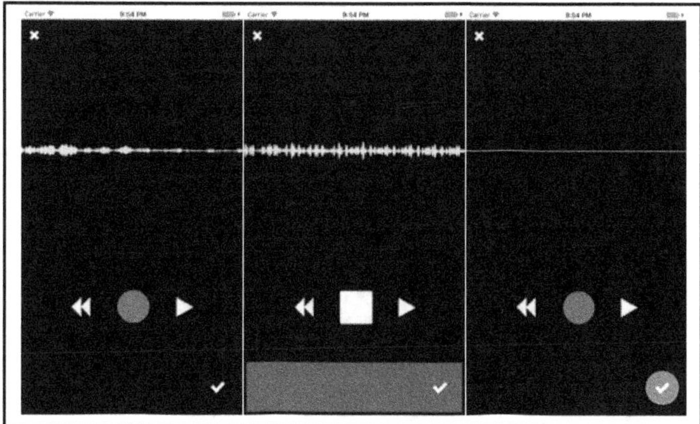

Finishing touches

Let's use color to finalize our initial app style. It's time to change the base color used in the `ActionBar` to provide the overall feeling we want with the app. Let's start by defining a few variables in `app/style/_variables.scss`:

```
// baseline theme colors
@import '~nativescript-theme-core/scss/dark';

$slate: #150e0c;

// page
$background: $black;
// action-bar
$ab-background: $black;
```

With those few changes, we have given our app a different (objectively sleeker) vibe:

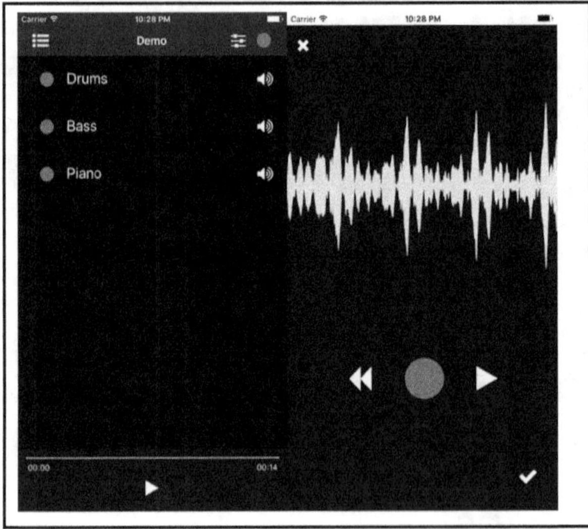

Summary

In this chapter, we were finally able to add some nice polishing touches to the app's look and feel. We were able to install the `nativescript-dev-sass` plugin, which adds a compilation step to build our CSS while maintaining a clean approach to styling. Knowing how best to take advantage of the core theme's SASS with proper file organization is key to gaining a flexible base to work with. Take the concepts presented in this chapter and let us know how they helped you achieve the styling goals you are after; we would love to hear about it!

We also took a look at how to work with the `nativescript-ngx-fonticon` plugin to utilize font icons throughout our app. This helped clean up clunky textual labels with concise iconic visuals.

In the next chapter, we will take a look at how to unit test several key features to future proof our app's codebase against new feature integrations, which might introduce regressions. Testing to the rescue!

12
Unit Testing

Let's start this chapter with testing; most people think testing is boring. Guess what, they are mostly right! Testing can be fun in that you get to try and break your code, but it can sometimes be tedious work. However, it can help you catch mistakes before your customers do, and as a bonus, it can prevent you from making the same bug occur multiple times. How much is your reputation to your clients or customers worth? A little tedious work could mean the difference between a Triple-A application, and a mediocre application.

In this chapter, we will cover the following topics:

- Angular Testing Framework
- NativeScript Testing Framework
- How to write tests using Jasmine
- How to run Karma tests

Unit testing

Unit tests are used to test small pieces of application code functionality for correctness. This also allows us to verify that the functionality continues to work as expected, when you refactor code and/or add new features. Both NativeScript and Angular offer unit testing frameworks. We will explore both types of unit testing, as they both have pros and cons.

Developing tests at any time is good. However, it is preferable to develop them alongside your development of the project's code. Your mind will be fresh on the new features, modifications, and all the new code you just added. In our case, because we are presenting lots of new concepts throughout this book, we haven't followed the best practice since it would have complicated the book even more. So, although adding tests later is good, adding them before or while adding your new code is considered the best practice.

Angular testing

The first type of unit testing we will cover is Angular unit testing. It is based around **Karma** (`https://karma-runner.github.io/`) and **Jasmine** (`http://github.com/pivotal/jasmine`). Karma is a full-featured test runner, which was produced by the Angular team. When the team was implementing Angular, they ran into issues , such as figuring out how to test Angular, so they built Karma. Karma has ended up becoming an industry-standard multipurpose test runner. Jasmine is an open source testing framework that implements a number of testing constructs helps you do all your testing easily. It has been around a lot longer than Karma. Since it was in use by a lot of people before Karma, it became the default testing library in the Angular community. You are free to use any of the other frameworks, such as Mocha, Chia, or even your own home-grown testing framework. However, since almost everything you will see in the Angular community is based around Jasmine, we will use it also.

Let's get the pieces you need installed for Angular testing in NativeScript:

```
npm install jasmine-core karma karma-jasmine karma-chrome-launcher --save-
dev
npm install @types/jasmine karma-browserify browserify watchify --save-dev
```

You also should probably install Karma globally, especially on Windows. However, it is helpful to do so on other platforms also so that you can just type `karma` and it runs. In order to do that, type the following command:

```
npm -g install karma
```

If you don't have TypeScript installed globally, where you can just type `tsc`, and it builds, you should install it globally. You have to transpile your TypeScript into JavaScript before running any tests. To install TypeScript globally, type the following command:

```
npm -g install typescript
```

Karma was designed to run tests inside a browser; however, NativeScript code does not run in a browser at all. So, we have to do a couple of things differently to make the standard Karma testing system run with some NativeScript application code. The normal Angular-specific Karma configuration won't work in most cases. If you are going to do any Angular work with the web side of things, you should check out the standard Angular testing QuickStart project (`https://github.com/angular/quickstart/`). That project will get everything set up for a traditional Angular application that runs in a browser.

However, in our case, because we are using NativeScript Angular, we will need a totally customized `Karma.conf.js` file. We have included it in the custom config file in the git repo, or you can type it from here. Save this file as `Karma.ang.conf.js`. We are giving a different configuration name because the NativeScript testing we discuss later will use the default `Karma.conf.js` name:

```
module.exports = function(config) {
   config.set({
     // Enable Jasmine (Testing)
     frameworks: ['jasmine', 'browserify'],

     plugins: [
       require('karma-jasmine'),
       require('karma-chrome-launcher'),
       require('karma-browserify')
     ],

     files: [ 'app/**/*.spec.js' ],

     preprocessors: {
        'app/**/*.js': ['browserify']
     },

     reporters: ['progress'],

     browsers: ['Chrome'],
  });
};
```

This configuration sets it up so that Karma will use Jasmine, Browserify, and Chrome to run all the tests. Since Karma and Angular were designed for browsers first, all the testing still has to run inside a browser. This is the major downside for the Angular testing system when doing NativeScript code. It won't support any NativeScript-specific code. So, this type of testing is better done on files that are data models and/or any code that has no NativeScript-specific code in it, which unfortunately in some of your apps might not be very much code. However, if you are doing both a NativeScript and web application using the same code base, then you should have a lot of code that can run through the standard Angular testing framework.

For Angular testing, you will create Jasmine specification files, and they all end with `.spec.ts`. We must create these files in the same directory as the code you are testing. So, let's take a crack at creating a new specification file for testing. Since this type of unit testing does not allow you to use any NativeScript code, I chose a random model file to show how easy this type of unit testing is. Let's create a file called `track.model.spec.ts` in the `app/modules/shared/models` folder; this file will be used to test the `track.model.ts` file in that same folder. Here is our test code:

```
// This disables a issue in TypeScript 2.2+ that affects testing
// So this line is highly recommend to be added to all .spec.ts files
export = 0;

// Import our model file (This is what we are going to test)
// You can import ANY files you need
import {TrackModel} from './track.model';

// We use describe to describe what this test set is going to be
// You can have multiple describes in a testing file.
describe('app/modules/shared/models/TrackModel', () => {
  // Define whatever variables you need
  let trackModel: TrackModel;

  // This runs before each "it" function runs, so we can
  // configure anything we need to for the actual test
  // There is an afterEach for running code after each test
  // If you need tear down code
  beforeEach( () => {
    // Create a new TrackModel class
    trackModel = new TrackModel({id: 1,
       filepath: 'Somewhere',
       name: 'in Cyberspace',
       order: 10,
       volume: 5,
       mute: false,
       model: 'My Model'});
  });

  // Lets run the first test. It makes sure our model is allocated
  // the beforeEach ran before this test, meaning it is defined.
  // This is a good test to make sure everything is working properly.
  it( "Model is defined", () => {
    expect(trackModel).toBeDefined();
  });

  // Make sure that the values we get OUT of the model actually
  // match what default values we put in to the model
  it ("Model to be configured correctly", () => {
```

```
      expect(trackModel.id).toBe(1);
      expect(trackModel.filepath).toBe('Somewhere' );
      expect(trackModel.name).toBe('in Cyberspace');
      expect(trackModel.order).toBe(10);
      expect(trackModel.model).toBe('My Model');
    });

    // Verify that the mute functionality actually works
    it ('Verify mute', () => {
      trackModel.mute = true;
      expect(trackModel.mute).toBe(true);
      expect(trackModel.volume).toBe(0);
      trackModel.mute = false;
      expect(trackModel.volume).toBe(5);
    });

    // Verify the volume functionality actually works
    it ('Verify Volume', () => {
      trackModel.mute = true;
      expect(trackModel.volume).toBe(0);
      trackModel.volume = 6;
      expect(trackModel.volume).toBe(6);
      expect(trackModel.mute).toBe(false);
    });
  });
```

So, let's break this down. The first line fixes an issue with testing inside a browser with a TypeScript-built file that uses modules. As I noted in the comments, this should be added to all your `spec.ts` files. The next line is where we load our model that we will test; you can import any files that you need here, including the Angular library.

Remember that a `.spec.js` file is just a normal TypeScript file; the only thing that differentiates it is that it has access to Jasmine globalized functions, and runs in a browser. So, all your normal TypeScript code will work fine.

The following line is where we start the actual testing framework. It is a Jasmine function that is used to create a test. Jasmine uses a `describe` function to start a group of tests. Describe has two parameters: the text description to print, and then the actual function to run. So, we basically put in the name of the model we are testing and then create the function. Inside each `describe` function, we add as many `it` functions as we need. Each `it` is used for a subset of the tests. You can also have multiple `describes`, if necessary.

So, in our test here, we have four separate test groups; the first one is really just to verify that everything is correct. It simply makes sure that our model gets defined properly. So, we are just using the Jasmine `expect` command to test for a valid object that was created with the `.toBeDefined()` function. Simple, isn't it?

The next test set attempts to make sure that the defaults are set properly from the `beforeEach` function. As you can see, we are using the `expect` command again with the `.toBe(value)` function. This is actually highly recommended; it seems obvious that the values set should match the values read, but you want to treat your modules as a black box. Verify all input and output to make sure that it really is set the same way you set it. So, even though we know we set ID to 1, we are verifying that when we get the ID, it still equals 1.

The third test function starts testing the mute capability, and the final one tests the volume functionality. Note that both the *mute* and *volume* have several states and/or impact multiple variables. Anything that is beyond a simple assignment should be tested through every single state you know of, both valid and invalid, if possible. In our case, we noted that mute affects volume and vice versa. So, we verify that when one has changed, the other changes with it. This is used as a contract to make sure that, down the road if this class changes, it will remain the same externally, or our tests will break. In this case, this is more of a brown-box; we know a side-effect of mute, and we are depending on that side-effect in the application, so we will test for that side-effect to make sure that it never changes.

Running the tests

Now, let's run the test by typing `tsc` to create the transpiled JS files, and then running the following command:

```
karma start karma.ang.conf.js
```

Karma will then find ALL of the `.spec.js` files and then run all of them on your Chrome browser, testing all the functionalities that you defined in each `.spec.js` file.

Unexpected test failure

```
10 07 2017 03:32:54.901:INFO [framework.browserify]: registering rebuild (autoWatch=true)
10 07 2017 03:32:56.831:INFO [framework.browserify]: 42677 bytes written (0.86 seconds)
10 07 2017 03:32:56.835:INFO [framework.browserify]: bundle built
10 07 2017 03:32:56.844:WARN [karma]: No captured browser, open http://localhost:9876/
10 07 2017 03:32:56.916:INFO [karma]: Karma v1.7.0 server started at http://0.0.0.0:9876/
10 07 2017 03:32:56.918:INFO [launcher]: Launching browser Chrome with unlimited concurrency
10 07 2017 03:32:56.939:INFO [launcher]: Starting browser Chrome
10 07 2017 03:32:59.286:INFO [Chrome 56.0.2924 (Windows 7 0.0.0)]: Connected on socket 2DaYeY_GPNQyo
6hbAAAA with id 43527929
LOG: 'setting volume from TrackModel:', 5
LOG: 'setting mute from TrackModel:', false
LOG: 'setting volume from TrackModel:', 1
LOG: 'setting volume from TrackModel:', 5
LOG: 'setting mute from TrackModel:', false
LOG: 'setting volume from TrackModel:', 1
LOG: 'setting volume from TrackModel:', 5
LOG: 'setting mute from TrackModel:', false
LOG: 'setting volume from TrackModel:', 1
LOG: 'setting mute from TrackModel:', true
LOG: 'setting volume from TrackModel:', 0
LOG: 'setting mute from TrackModel:', false
LOG: 'setting volume from TrackModel:', 1
LOG: 'setting volume from TrackModel:', 5
LOG: 'setting mute from TrackModel:', false
LOG: 'setting volume from TrackModel:', 1
LOG: 'setting mute from TrackModel:', true
LOG: 'setting volume from TrackModel:', 0
LOG: 'setting volume from TrackModel:', 6
Chrome 56.0.2924 (Windows 7 0.0.0) TrackModel Creation Verify mute FAILED
        Expected 1 to be 5.
            at Object.<anonymous> (C:/Users/NATHAN~1/AppData/Local/Temp/3f19dd2ffb15b09069e4667c0934
bc50.browserify:89:35)
Chrome 56.0.2924 (Windows 7 0.0.0): Executed 4 of 4 (1 FAILED) (0.094 secs / 0.018 secs)
```

Now isn't it very interesting that one of our tests actually failed; `TrackModel Creation Verify mute FAILED` and `Expected 1 to be 5.`. This failure was not preplanned for the book; it is actually a real corner case that we just found because we started to use unit testing. If you want to take a quick look at the code, here is the `TrackModel.ts` code stripped down to just show the relevant routines:

```
export class TrackModel implements ITrack {
 private _volume: number = 1;
 private _mute: boolean;
 private _origVolume: number;
 constructor(model?: ITrack) {
   if (model) {
    for (let key in model) {
      this[key] = model[key];
    }
   }
 }

 public set mute(value: boolean) {
   value = typeof value === 'undefined' ? false : value;
   this._mute = value;
```

```
    if (this._mute) {
      this._origVolume = this._volume;
      this.volume = 0;
    } else {
      this.volume = this._origVolume;
    }
  }

  public set volume(value: number) {
    value = typeof value === 'undefined' ? 1 : value;
    this._volume = value;
    if (this._volume > 0 && this._mute) {
      this._origVolume = this._volume;
      this._mute = false;
    }
  }
}
```

Now, I'll give you a few minutes to look at the preceding test code and this code and check whether you can spot why the test fails.

Good, I see, you are back; did you see where the corner case is? Don't feel bad if you couldn't find it quickly; it took me a few minutes to figure out why it was failing also.

Well, first, look at the error message; it said `Verify Mute FAILED`, so this means it was our mute test that failed. Then, we put `Verify mute` in the `it` function that tested our mute functionality. The second clue is the error, `Expected 1 to be 5`. So, we expected something to be 5, but it was actually 1. So, this specific test and this line of code are failing in the test:

```
it ('Verify mute', () => {
    expect(trackModel.volume).toBe(5);
});
```

Why did it fail?

Let's start by looking at the test initialization, `beforeEach`; you will see that `` `mute: false` ``. Well, next, let's look at the constructor; it basically does `this.mute = false` and the mute setter then runs down its `else` path, which is `this.volume = this._origVolume`. Guess what? `this._origVolume` has not been set yet, so it sets `this.volume = undefined`. Now take a look at the volume routine; the new volume comes in `undefined`, it is set to `1`, which overrides our original setting of 5. So, the test `Expected 1 to be 5.` fails.

Interesting corner case; it wouldn't have happened if we hadn't set `mute` to `false` as part of testing the initialization of properties. However this is something we should test, because maybe in one of the revisions of the application we will store the mute value and restore it on start.

To fix this, we should fix the class a little. We'll let you do the changes that you think are necessary to fix this issue. If you get stuck, you can rename the `track.model.fixed.ts` based on the `track.model.ts` file; it contains the correct code.

Once you have it fixed, run the same `tsc` and then the `karma start karma.ang.conf.js` command; you should see everything is successfull.

Test passes

```
10 07 2017 04:35:32.865:INFO [framework.browserify]: registering rebuild (autoWatch=true)
10 07 2017 04:35:34.380:INFO [framework.browserify]: 42788 bytes written (0.55 seconds)
10 07 2017 04:35:34.386:INFO [framework.browserify]: bundle built
10 07 2017 04:35:34.392:WARN [karma]: No captured browser, open http://localhost:9876/
10 07 2017 04:35:34.465:INFO [karma]: Karma v1.7.0 server started at http://0.0.0.0:9876/
10 07 2017 04:35:34.468:INFO [launcher]: Launching browser Chrome with unlimited concurrency
10 07 2017 04:35:34.541:INFO [launcher]: Starting browser Chrome
10 07 2017 04:35:36.685:INFO [Chrome 56.0.2924 (Windows 7 0.0.0)]: Connected on socket Gw52EZtmuNPDI
g8BAAAA with id 409218
LOG: 'setting volume from TrackModel:', 5
LOG: 'setting volume from TrackModel:', 5
LOG: 'setting volume from TrackModel:', 5
LOG: 'setting mute from TrackModel:', true
LOG: 'setting volume from TrackModel:', 0
LOG: 'setting mute from TrackModel:', false
LOG: 'setting volume from TrackModel:', 5
LOG: 'setting volume from TrackModel:', 5
LOG: 'setting mute from TrackModel:', true
LOG: 'setting volume from TrackModel:', 0
LOG: 'setting volume from TrackModel:', 6
Chrome 56.0.2924 (Windows 7 0.0.0): Executed 4 of 4 SUCCESS (0.061 secs / 0.015 secs)
```

As this example pinpointed, your code might run correctly in some cases, but fail in others. Unit testing can pinpoint errors in your logic that you might not see readily. This is especially important when adding new features and/or fixing bugs. It is highly recommended that you create new tests for both, and then you will know at least at a minimum that your new or modified code is behaving properly after you make any code changes.

Let's switch gears slightly, and look at the NativeScript testing framework; the Angular framework is pretty cool, but it has the nasty limitation of not having NativeScript framework calls available, so it limits a lot of its usefulness.

NativeScript testing framework

Okay, be ready to play with the NativeScript testing framework. It is fairly simple to install, you simply type the following command:

```
tns test init
```

There is no reason to switch testing frameworks, so choose `jasmine` at the prompt asking you which testing framework to use with the NativeScript testing framework. This will install all the needed resources for the NativeScript testing system. NativeScript's testing system also uses Karma, and it supports a couple of different testing frameworks, but for consistency, we will want to continue to use Jasmine.

Remember earlier when I said Karma uses a browser to do all its tests, and I also said that NativeScript code doesn't run in a browser? So, why does NativeScript use Karma? How does Karma run the NativeScript code? Excellent questions! Karma is actually tricked into thinking that your NativeScript application is a browser. Karma uploads the tests to the browser (that is, the NativeScript application), and it then runs them. So, in effect, your application is a browser to Karma; this is a pretty ingenious solution by the NativeScript team.

Now, the big pro of the NativeScript testing system is that it can actually test all your NativeScript code. It will automatically run a special build of your application in your emulator (or on a real device) so that it can run all the NativeScript code and access the device properly. The biggest con of the NativeScript testing system is that it requires a lot more resources because it must use an emulator (or real device) to run the tests. So, it can be considerably more time-consuming to run testing than the standard unit testing we discussed earlier in this chapter.

Okay, so now you have it all installed. Let's move on. All of the NativeScript testing files will be in the `app/tests` folder. This folder was created when you ran the `tns test init`. If you open up that folder, you will see `example.js`. Feel free to delete or leave the file. It is just a dummy test to show you how to format your tests using Jasmine.

So, for our NativeScript test, I selected a simple service that uses NativeScript code. Let's create our `database.service.test.ts` file in the `app/test` folder. Your files in this folder can be named anything but, to keep things easier to find, we will end it with `.test.ts`. You can also create subdirectories to organize all your tests. In this case, we will test the `app/modules/core/services/database.service.ts` file.

This specific service, if you look at the code, actually uses the NativeScript `AppSettings` module to store and retrieve data from the Android and iOS system-wide storage system. So, this is a great file to test. Let's create our test file:

```
// Import the reflect-metadata because angular needs it, even if we don't.
// We could import the entire angular library; but for unit-testing;
// smaller is better and faster.
import 'reflect-metadata';

// Import our DatabaseService, we need at least something to test... ;-)
import { DatabaseService } from
"../modules/core/services/database.service";

// We do the exact same thing as we discussed earlier;
// we describe what test group we are testing.
describe("database.service.test", function() {

  // So that we can easily change the Testing key in case we find out later
in our app
  // we need "TestingKey" for some obscure reason.
  const TestingKey = "TestingKey";

  // As before, we define a "it" function to define a test group
  it("Test Database service class", function() {

    // We are just going to create the DatabaseService class here,
    // no need for a beforeEach.
    const dbService = new DatabaseService();

    // Lets attempt to write some data.
    dbService.setItem(TestingKey, {key: "alpha", beta: "cygnus", delta:
true});

    // Lets get that data back out...
    let valueOut = dbService.getItem(TestingKey);

    // Does it match?
    expect(valueOut).toBeDefined();
    expect(valueOut.key).toBe("alpha");
    expect(valueOut.beta).toBe("cygnus");
    expect(valueOut.delta).toBe(true);

    // Lets write some new data over the same key
    dbService.setItem(TestingKey, {key: "beta", beta: true});

    // Lets get the new data
    valueOut = dbService.getItem(TestingKey);
```

```
        // Does it match?
        expect(valueOut).toBeDefined();
        expect(valueOut.key).toBe("beta");
        expect(valueOut.beta).toBe(true);
        expect(Object.keys(valueOut).length).toBe(2);

        // Lets remove the key
        dbService.removeItem(TestingKey);

        // Lets make sure the key is gone
        valueOut = dbService.getItem(TestingKey);
        expect(valueOut).toBeFalsy();
    });
});
```

You might already be able to read this test file pretty easily. Basically, it calls the database service a couple of times to set the same key with different values. Then, it asks the database service to return the values stored and verifies that the results match what we stored. Then, we tell the database service to delete our storage key and verify that the key is gone, all pretty straightforward. The only thing different in this file is the `include 'reflect-metadata'`. This is because the database service uses metadata in it, so we have to make sure that the metadata class is loaded before we load the database service class.

Running the tests

Let's try testing the application; to run your tests, type the following command:

```
tns test android
```

Alternatively, you can run the following command:

```
tns test ios
```

This will start the testing, and you should see something like this:

```
JS: NSUTR: successfully connected to karma
10 07 2017 06:11:23.743:INFO [NativeScript / 19 (4.4.4; Samsung Galaxy S5 - 4.4.4 - API 19)]: Connec
ted on socket CBHPvInKrGh_Ej4aAAAI with id NativeScriptUnit-6452
JS: NSUTR: downloading http://192.168.32.1:9876/context.json
JS: NSUTR: eval script /base/node_modules/jasmine-core/lib/jasmine-core/jasmine.js?da99c5b057693d025
fad3d7685e1590600ca376d
JS: NSUTR: eval script /base/node_modules/karma-jasmine/lib/boot.js?945a38bf4e45ad2770eb94868231905a
04a0bd3e
JS: NSUTR: eval script /base/node_modules/karma-jasmine/lib/adapter.js?7a813cc290d592e664331c573a1a7
96192cdd1ad
JS: NSUTR: require script /base/app/tests/database.service.test.js?3a4b41cd376983c3dc219481ed45d9273
50647b0 from ../../tests/database.service.test.js
JS: NSUTR: require script /base/app/tests/example.js?d4cad4cd203df70666c986cd09ce402b21d89b39 from .
./../tests/example.js
JS: NSUTR: beginning test run
NativeScript / 19 (4.4.4; Samsung Galaxy S5 - 4.4.4 - API 19): Executed 0 of 2 SUCCESS (0 secs / 0 s
NativeScript / 19 (4.4.4; Samsung Galaxy S5 - 4.4.4 - API 19): Executed 1 of 2 SUCCESS (0 secs / 0.0
NativeScript / 19 (4.4.4; Samsung Galaxy S5 - 4.4.4 - API 19): Executed 2 of 2 SUCCESS (0 secs / 0.0
85 secs)
NativeScript / 19 (4.4.4; Samsung Galaxy S5 - 4.4.4 - API 19): Executed 2 of 2 SUCCESS (0.162 secs /
 0.085 secs)
JS: NSUTR: completeAck
10 07 2017 06:11:25.529:WARN [NativeScript / 19 (4.4.4; Samsung Galaxy S5 - 4.4.4 - API 19)]: Discon
NativeScript / 19 (4.4.4; Samsung Galaxy S5 - 4.4.4 - API 19) ERROR
  Disconnectedundefined
NativeScript / 19 (4.4.4; Samsung Galaxy S5 - 4.4.4 - API 19): Executed 2 of 2 SUCCESS (0.162 secs /
 0.085 secs)
```

Note that there is an `ERROR` on this screen; this is a false error. Basically, when the app finishes running its tests, it quits. Karma sees that the application has quit unexpectedly and logs it as an `"ERROR" Disconnected`. The import information is the line below the error, where it says `Executed 2 of 2 SUCCESS`. This means that it ran two different `described` tests (that is, our `test.ts` file and the extra `example.js` file).

You might have also noted that our test file is identical to the Angular testing file. That is because they both use Jasmine and Karma. So, the test files can be set up almost identically. In this specific case, because the testing is actually running inside your application, any plugins, code, and modules, including any native code are all available to be tested. This is what makes the NativeScript testing harness a lot more powerful and useful. However, its greatest strength is also its weakness. Since it has to run inside a running NativeScript application, it takes a lot more time to build, start, and run all the tests. This is where the standard Angular testing framework can prove beneficial over the NativeScript testing framework. Anything that is not using any NativeScript-specific code can run from your command line almost instantly, with very little overhead. The quicker your tests run, the more likely you are to run them frequently.

Summary

In this chapter, we discussed how to do unit tests and the pros and cons of two of the methods of doing unit tests. In a nutshell, Angular testing works for generic TypeScript code that does not call any NativeScript-specific code, and it runs your tests really quickly. The NativeScript testing harness runs inside your NativeScript application and has full access to anything you write and anything a normal NativeScript application can do. However, it requires the NativeScript application to be running to run its tests, so it might require a full build step before it can run your tests.

Now that we have discussed the two types of unit testing, hang on to your testing hat. In the next chapter, we will cover how to do end-to-end testing or full screen and application testing of your awesome application.

13
Integration Testing with Appium

In the preceding chapter, we explored how to do unit testing, but unit testing doesn't allow you to test whether the button used in your app still actually runs a function, or what happens when the user swipes left. For that, we will need application testing or end-to-end testing. Well, let's start learning end-to-end testing; this is where testing can get complicated and fun.

In this chapter, we will cover the following topics:

- Appium testing framework
- Writing MochaJS, ChaiJS, and ShouldJS tests
- How to find and interact with elements on the screen
- How to run the tests
- Travis and GitHub integration

Integration testing

There are several full application frameworks out there, but we will show you how to use Appium (`http://appium.io`). Appium is an awesome open source application testing framework. Appium supports both iOS and Android, which makes it a perfect fit for doing all our on-device testing. You want to start creating tests to test your basic flow through your application, and even create more complicated tests that test alternate flows through your app.

Let's get it installed first; run the following command:

```
npm install appium wd nativescript-dev-appium --save-dev
```

The preceding command installs Appium, the Appium communication driver **WD** (`http://admc.io/wd/`), and the **NativeScript driver** (`https://github.com/NativeScript/nativescript-dev-appium`). The WD driver is what communicates with Appium and the NativeScript driver. The `nativescript-dev-appium` is the driver that interacts with WD and your test code. In reality, the NativeScript driver is a very thin wrapper around the WD driver and just facilitates some configuration and then exposes the WD driver to your application. So interaction commands will be found in the WD documentation.

Application/Integration testing takes a bit more work, as you have to programmatically run it as a normal user would interact with your app. So, you have to do things, such as find the button element, then perform `button.tap()`. So, your tests might be a bit verbose, but this allows you to test any, and all, functionalities. The down side is that this is a lot more time-consuming to run and takes more maintenance work when you change screens. However, the payoff is that when you add code, it verifies that your app still runs properly on each screen automatically, and you can test it on multiple devices and resolutions, once again automatically.

After installation, you will have a brand new `e2e-tests` folder in your root folder of your application. This folder is where all your end-to-end test files will live. Now, one thing that you need to know is that the Appium NativeScript driver uses the MochaJS test framework (`https://mochajs.org/`). The Mocha testing framework is similar to the Jasmine framework, which we discussed in the preceding chapter. It uses the same `describe` and `it` functions for the start of the tests, just like Jasmine. In addition, it uses the Chai (`http://chaijs.com/`) and ShouldJS (`https://github.com/shouldjs/should.js`) testing frameworks that work hand in hand with the Mocha test framework and the WD driver.

Another thing to note is that all these are designed around pure JavaScript. You can get typings for Mocha, Should, and Chai, but for the NativeScript Appium driver or WD driver, typings don't exist. You can use TypeScript, but it is awkward, as commands are not just WD-based commands but chained through mocha. TypeScript gets easily confused as to what context you are in. So, mostly the Appium tests are created in pure JavaScript rather than in TypeScript. However, feel free to use TypeScript if you want; just make sure that you run `tsc` to build the `JS` files before you run the tests.

Configuration

One more setup step that you will need to do is to create an `appium.capabilities.json` file in the root folder of your project. This is basically a configuration file that you can use to configure the emulator that you need to run any of your tests on. The file is documented on the Appium site but to get you up and running quickly you can use the paired-down file which we use as follows:

```
{
 "android44": {
   "browserName": "",
   "appium-version": "1.6.5",
   "platformName": "Android",
   "platformVersion": "4.4",
   "deviceName": "Android 44 Emulator",
   "noReset": false,
   "app": ""
 },

 "ios10phone": {
   "browserName": "",
   "appium-version": "1.6.5",
   "platformName": "iOS",
   "platformVersion": "10.0",
   "deviceName": "iPhone 6 Simulator",
   "app": ""
 }
}
```

We've simplified it and removed all other emulator entries to save space. However you give each emulator entry a key-- you tell Appium using the key the emulator configuration that you will be running. This example file shows two configurations. The first one is an Android 4.4 device, and the second one is an iOS simulator (iPhone 6 runs iOS 10). You can have as many configurations as you want in this file. When you run Appium, you can tell it which device you will target, using the `--runType=KEY` parameter.

Creating a test

Let's start our journey and create a new test file: `list.test.js`. This file will test our mix-list screen. The screen's HTML (`/app/modules/mixer/components/mix-list.component.html`) looks like this:

```
<ActionBar title="Compositions" class="action-bar">
   <ActionItem (tap)="add()" ios.position="right">
```

```
      <Button [text]="'fa-plus' | fonticon" class="fa action-item"></Button>
    </ActionItem>
  </ActionBar>
  <ListView [items]="(mixer$ | async)?.compositions | orderBy: 'order'"
  class="list-group">
   <ng-template let-composition="item">
    <GridLayout rows="auto" columns="100,*,auto" class="list-group-item">
      <Button [text]="'fa-pencil' | fonticon" (tap)="edit(composition)"
  row="0" col="0" class="fa"></Button>
      <Label [text]="composition.name" (tap)="select(composition)" row="0"
  col="1" class="h2"></Label>
      <Label [text]="composition.tracks.length" row="0" col="2" class="text-
  right"></Label>
    </GridLayout>
   </ng-template>
  </ListView>
```

We've included the code here so that you can see easily how we made the test with the details provided on the screen.

```
// In JavaScript code, "use strict"; is highly recommended,
// it enables JavaScript engine optimizations.
"use strict";

// Load the Appium driver, this driver sets up our connection to Appium
// and the emulator or device.
const nsAppium = require("nativescript-dev-appium");
```

We need to include the NativeScript Appium driver in our JavaScript test code; this is what is used to actually communicate and set up the Mocha, ShouldJS, WD, Appium, and Chia to be able to work properly. The following line alone is required for your usage:

```
// Just like Jasmine, Mocha uses describe to start a testing group.
describe("Simple example", function () {

 // This is fairly important, you need to give the driver time to wait
 // so that your app has time to start up on the emulator/device.
 // This number might still be too small if you have a slow machine.
 this.timeout(100000);
```

As the comments in the source code mention, it is critical that you give enough time to start up Appium and the application in your emulator. So, out personal default is `100,000`; you can play with this number, but this is the maximum amount of time it will wait before it will declare the tests a failure. Having a larger value means that you give your emulator and Appium more time to actually start running. Appium gives you its startup output quickly, but when it goes to actually initialize the test and driver, that process takes a lot of time. Once the test starts to run, it runs very fast:

```
 // This holds the driver; that will be used to communicate with Appium &
Device.
 let driver;

 // This is ran once before any tests are ran. (There is also a beforeEach)
 before(function () {
    // VERY, VERY important line here; you NEED a driver to communicate to
your device.
    // No driver, no tests will work.
    driver = nsAppium.createDriver();
 });
```

It is also very important to initialize and create the driver before your tests are run. This driver is global throughout your tests. So, we will declare it globally in the `describe` function and then initialize it with the Mocha `before` function that runs before any tests are run:

```
// This is ran once at the end of all the tests. (There is also a
afterEach)
after(function () {

  // Also important, the Appium system works off of promises
  // so you return the promise from the after function
  // NOTICE no ";", we are chaining to the next command.
  return driver

    // This tells the driver to quit....
    .quit()
    // And finally after it has quit we print it finished....
    .finally(function () {
       console.log("Driver quit successfully");
    });
 });
```

We also add a Mocha after function to shut down the driver when we are all done. It is very important to ensure that any time you are working with the driver that you properly return it. Almost every single test piece is actually a promise underneath it. If you forget to return the promise, the testing harness will get royally mixed up and may run tests out of order and even close the driver before the tests are completed. So, always return the promise:

```
// Just like jasmine, we define a test here.
it("should find the + button", function () {

  // Again, VERY important, you need to return the promise
  return driver

  // This searches for an element by the Dom path; so you can find sub
```

```
items.
  .elementByXPath("//" + nsAppium.getXPathElement('Button'))
```

The `it` function is used just like we did in Jasmine--you are describing a test you plan on running so that you can find it if the test fails. Again, we return the promise chain; it is very important that you don't forget to do this. The driver variable is what gives us the different functionality when dealing with the emulators. So, the documentation for the functionality is in the WD repository, but I will give you a quick overview to get you going.

`.elementByXPath` and `.elementById` are really the only two functions that work well to properly find elements with NativeScript . However there is also a `.waitForElementByXPath` and `.waitForElementById` which both wait for the elements to show up. If you look at the documentation, you will observe a lot of `elementByXXX` commands, but Appium was designed for a browser, and NativeScript is not a browser. That is why, only some commands that have been emulated in the nativescript-dev-appium driver work to find elements in the NativeScript DOM.

So our test says find an element by XPath. XPath allows you to drill into your DOM and find components any level deep and also subcomponents of other components. So, if you do something like `/GridLayout/StackLayout/Label`, it will find a `Label` that is a child of a `StackLayout`, which is a child of a `GridLayout`. Using // will mean that you can find that element at any level in the DOM. Finally the `nsAppium.getXPathElement` is a method which was added to the official NativeScript driver by Nathanael Anderson to allow us to make the XPath tests cross-platform. In all reality, what you are passing to the XPath function is the real native name of the object. For example, a button on Android is a `android.widget.Button` or it could be a `UIAButton` or a `XCUIElementTypeButton` on iOS. So because you don't want to hardcode `getByElementXPath("android.widget.Button")`, this helper function translates the NativeScript `Button` to the proper underlying OS element that NativeScript actually uses when it creates a button in NativeScript. If, in the future, you add a plugin that uses an element that the `getXPathElement` doesn't know about, you can still use the real name of the element for those tests:

```
        // This element should eventually exist
        .text().should.eventually.exist.equal('\uf067');
    });
```

`.text()` is a function that Appium driver exposes to get the text value of the elements it finds. The `.should.eventually.exist.equal` is a Mocha and Should code. We are basically making sure that once this item is found, it actually matches the Unicode value of F067, which in Font-Awesome is the Plus character (fa-plus). Once it exists, we are happy--the test either succeeds or fails, depending on whether we break the screen or the screen continues to be the way we expect it. In addition, after the **`.equal`**, we could have chained more commands, such as `.tap()`, to fire the button if we wanted.

Okay, let's look at the next test that runs:

```
it("should have a Demo label", function () {

  // Again, VERY important, you need to return the promise
  return driver

    // Find all Label elements, that has text of "Demo"
   .elementByXPath("//" + nsAppium.getXPathElement("Label") +
"[@text='Demo']")

   // This item should eventually exist
   .should.eventually.exist

   // Tap it
   .tap();
});
```

This test searches the screen to show the `Demo` ListView item. We are looking for a NativeScript Label (that is, `nsAppium.getXPathElement`) anywhere in the NativeScript DOM (that is, //) that contains the text value of Demo in it. (that is, `[@text='Demo']`). This element should eventually exist, and once it does exist, it calls the `tap()` function. Now, if you look at the source code, you will see the following:

```
<Label [text]="composition.name" (tap)="select(composition)" row="0"
col="1" class="h2"></Label>
```

So, it will run the select function when the `tap` is fired. The `select` function ends up ultimately loading the `/app/modules/player/components/track-list/track-list.component.html` file, which is used to display the composition of that mixer item on the screen.

All the tests are executed sequentially, and the state of the app is preserved from one test to another. This means that tests are not independent like we are used to when writing unit-tests.

The next test we will verify is that the `Demo` Label actually switches screens in the next test after we tap on it:

```
it("Should change to another screen", function () {

   // As usual return the promise chain...
   return driver

   // Find all Label elements, that has text of "Demo"
   .waitForElementByXPath("//" + nsAppium.getXPathElement("Label") +
"[@text='Drums']")

   // This item should eventually exist
   .should.eventually.exist.text();
 });
```

So, now that we are on a new screen, we will verify that the `ListView` contains a Label with the name of `Drums`. This test just verifies that the screen actually changed when we tapped on the `Demo` Label in the prior test. We could have verified the text value but really if it exists, we are good. So, let's look at the next test:

```
it("Should change mute button", function () {

  // Again, returning the promise
  return driver

  // Find all Label elements that contains the FA-Volume
  .waitForElementByXPath("//" + nsAppium.getXPathElement("Label") +
"[@text='\uf028']")

  // This item should eventually exist
  .should.eventually.exist

  // It exists, so tap it...
  .tap()

  // Make sure the text then becomes the muted volume symbol
  .text().should.eventually.become("\uf026");
});

// This closes the describe we opened at the top of this test set.
});
```

Our final example test shows chaining. We search the Label that has the volume control symbol. Then, once it exists, we tap on it. Then, we verify that the text actually became the volume off symbol. The `f028` is the Font Awesome Unicode value for `fa-volume-up`, and the `f026` is the Font Awesome Unicode value for `fa-volume-off`.

So now that you have this really cool test, you want to launch your emulator. The emulator should be already running. You also should ensure that you have the latest version of the application on the device. Then, to run the test, simply type the following command:

```
npm run appium --runType=android44
```

Ensure that you put in which run type configuration you are going to use, and you should see something like this after a few minutes:

```
> nativescript-dev-appium

Using project-local Appium binary.
Custom capabilities found.


  Simple example
Getting caps.app!
testRunType android1
Creating driver!
    √ should find the + button (24181ms)
    √ should have a Demo label (1091ms)
    √ Should change to another screen (2321ms)
    √ Should change mute button (1319ms)
Driver quit successfully


  4 passing (31s)

Test runner exited with code 0
```

Remember that Appium end-to-end tests take a while to start, so if it looks like it is frozen for a while, don't panic and quit it. It may take 24 seconds for the first test, seconds for each additional test. The first test has all the time in it. It is normal for Appium to take a long time to start the driver and the application on the emulator. This delay normally occurs after you see the first couple of lines of text printed, as shown in the preceding screen, so, have some patience.

More Appium testing

I wanted to include one more test (not used in this application) that I have written in the past for a different project since this will give you an idea of just how powerful Appium can be:

```
it("should type in an element", function (done) {
  driver
  .elementByXPath('//' + nsAppium.getXPathElement("EditText") +
"[@text='Enter your name']")
  .sendKeys('Testing')
```

```
      .text()
      .then(function (v) {
         if ('Testing' !== v) {
            done(new Error("Value in name field does not match"));
         } else {
            done();
         }
      }, done);
   });
});
```

The first thing you might note is that I did not return the promise chain. That is because this example shows how to use the asynchronous support of `it`. For an async support, you can use a promise or make the function coming into `it` have a `done` callback function. When Mocha detects a callback function in the `it`, it will run your `it` tests in the async mode and doesn't need the promise to let it know that it can resume with the next test. Sometimes, you may just want to maintain full control or you may be calling code that requires async callbacks.

This test looks for a `EditText` element that contains `Enter your name`. Then, it actually types *Testing* into it using the `sendKeys`. Next, it asks for the `text` out of the field and uses the `then` part of the promise to check the value against the hardcoded testing. When it is all finished, it calls the `done` function. If you pass the `done` function an `Error` object, then it knows that the test failed. So, you can see in the `if` statement that we passed a `new Error` and that we put the `done` function in the `catch` part of the `then` statement.

We have barely scratched the surface of what you can do with Appium, Should, Mocha, and Chia. You can control pretty much all aspects of the application as if you were manually doing each step. Initially, in your development, manually testing is a lot faster. However, as you start to build up end-to-end tests, each time you make changes, you can check whether the app still works properly, and you do not have to sit in front of multiple devices for any amount of time--you just start the tests and see the results later.

Automated testing

One more thing that you should note is that the more automated you make your testing, the more likely you are to use it and gain from its benefits. If you constantly have to manually run the test, you are likely to get annoyed and stop running them. So automating this is critical in our opinion. Since there are many books written on this subject we are just going to give you a couple of pointers that you can research and then move forward.

Most source control systems allow you to create hooks. With these hooks, you can create a commit hook so that on check-in of any new code, your testing frameworks will run. These hooks are normally pretty simple to create as they are simple scripts that just run each time a commit is made.

In addition, if you are using GitHub, there are sites such as Travis that you can tie into easily without having to do any hook changes.

GitHub and Travis integration

Here is how you can do some integration with GitHub and Travis; this will allow your NativeScript Testing framework, which we discussed in the preceding chapter, to automatically run your tests on each change or pull request. Create a new `.travis.yml` file in the very root of your GitHub repository. This file should look like this:

```
language: android

jdk: oraclejdk8

android:
 components:
 - tools
 - platform-tools
 - build-tools-25.0.2
 - android-25
 - extra-android-m2repository
 - sys-img-armeabi-v7a-android-21

before_cache:
 - rm -f $HOME/.gradle/caches/modules-2/modules-2.lock

cache:
 directories:
 - .nvm
 - $HOME/.gradle/caches/
 - $HOME/.gradle/wrapper/

install:
 - nvm install node
 - npm install -g nativescript
 - tns usage-reporting disable
 - tns error-reporting disable

before_script:
 - echo no | android create avd --force -n test -t android-21 -b armeabi-
v7a
```

```
  - emulator -avd test -no-audio -no-window &
  - android-wait-for-emulator

script:
  - npm run travissetup
  - npm run travistest
```

This basically configures Travis to start an Android emulator; it waits for the emulator to start and then runs the `npm` commands. You can learn what these `npm` commands do from your `package.json`.

So, in your root application, that is, the package.json file of your app, you want to add the following keys:

```
"scripts": {
   "travissetup": "npm i && tns platform add android && tns build android",
   "travistest": "tns test android"
}
```

With these two changes, Travis will automatically test every single pull request to your repository, which then means that you can code, and Travis will continually do all your unit testing.

In addition, you can change the preceding Travis configuration file to add Appium to be installed and run also just by doing the following:

- Adding the Appium dependencies to your main `package.json` dependencies.
- Adding to the root of your project a `appium.capabilities.json` that has a `travisAndroid` key.
- Adding the `&& npm run appium --runType=travisAndroid` to your `travistest` key in the `package.json` file.

GitHub already has the integration with Travis built-in, so it is simple to document and get it running. If you use Gitlabs, you can use the Gitlabs CI system to do testing. In addition, you can use the repository hooks to use a wide number of other continuous integration services that are available. Finally, you can also develop your own.

Summary

In this chapter, we covered how to install and run Appium, how to build complete end-to-end tests and how to use the testing frameworks to test your screens fully. In addition, we covered how important it is to automate the running of the unit testing and Appium, and one way you can do so is using Travis and GitHub.

Now hang on tight--we will make a quick turn and start discussing how to deploy and use Webpack to optimize your builds for release.

14
Deployment Preparation with webpack Bundling

We want to deploy our app to the two leading mobile app stores, Apple App Store and Google Play; however, there are several things that we need to do to prepare our app for distribution.

To ensure that you use the smallest JavaScript size in addition to Angular's AoT compiler to help our app execute as fast as possible, we will use webpack to bundle everything. It's worth noting that webpack is not a requirement to create a distributable NativeScript app. However, it provides very nice benefits that should make it an important step for anyone when distributing their apps.

In this chapter, we will cover the following topics:

- Installing webpack for a NativeScript for Angular project
- Preparing a project to be bundled with webpack
- Solving various webpack bundling issues
- A primer on writing your own custom webpack plugin to solve specific cases

Using webpack to bundle the app

If it weren't for Sean Larkin, you might have never heard of webpack. His contributions and involvement in the bundler community have helped bring webpack into the Angular CLI and also make it a primary *go-to* bundler for many things. We greatly appreciate his efforts and kindness in the community.

Preparing to use webpack

Let's take a look at how webpack can be utilized to reduce the packaged size of our NativeScript for Angular app in addition to ensuring that it executes optimally on a user's mobile device.

Let's first install the plugin:

```
npm install nativescript-dev-webpack --save-dev
```

This automatically creates a `webpack.config.js` file (at root of project) preconfigured with a basic setup that will get you reasonably further with most apps. Additionally, it creates a `tsconfig.aot.json` file (also at root of project) since NativeScript's webpack usage will use Angular's AoT compiler while bundling. It also adds some nifty npm scripts to our `package.json` to help handle all the various bundling options we will want; consider the following example:

- `npm run build-android-bundle` to build for Android
- `npm run build-ios-bundle` to build for iOS
- `npm run start-android-bundle` to run on Android
- `npm run start-ios-bundle` to run on iOS

However, before we attempt those new commands, we will need to audit our app for a couple of things.

We should start by ensuring that all NativeScript import paths are preceded with `tns-core-modules/[module]`; consider the following example:

```
BEFORE:
import { isIOS } from 'platform';
import { topmost } from 'ui/frame';
import * as app from 'application';

AFTER:
import { isIOS } from 'tns-core-modules/platform';
import { topmost } from 'tns-core-modules/ui/frame';
import * as app from 'tns-core-modules/application';
```

We'll go through our app and do this now. This works fine for development and production builds.

You might wonder, *Hey! Why did you even use the other form if we needed to go through the entire codebase and change the imports after the fact?*

Great concern! There's actually a ton of examples out there that show the convenient shortform import path, so we chose to build the app using that throughout in this chapter to demonstrate that it works just fine for development to help avoid confusion, in case you come across examples such as these in the future. Besides, it doesn't take too much to edit that after the fact to prepare for webpack but now you know.

Run the following command right now:

```
npm run build-ios-bundle
```

We can see the following errors—which I have enumerated—and we will present solutions for it sequentially in the next section:

1. ERROR in Unexpected value `SlimSliderDirective` in `/path/to/TNSStudio/app/modules/player/directives/slider.directive.d.ts declared by the module PlayerModule in /path/to/TNSStudio/app/modules/player/player.module.ts`. Please add a `@Pipe/@Directive/@Component` annotation.
2. ERROR in Cannot determine the module for class `SlimSliderDirective` in `/path/to/TNSStudio/app/modules/player/directives/slider.directive.android.ts`! Add `SlimSliderDirective` to the `NgModule` to fix it. Cannot determine the module for class `SlimSliderDirective` in `/path/to/TNSStudio/app/modules/player/directives/slider.directive.ios.ts`! Add `SlimSliderDirective` to the `NgModule` to fix it.
3. ERROR in Error encountered resolving symbol values statically. Calling function `ModalDialogParams`, function calls are not supported. Consider replacing the function or lambda with a reference to an exported function, resolving symbol `RecorderModule` in `/path/to/TNSStudio/app/modules/recorder/recorder.module.ts`, resolving symbol `RecorderModule` in `/path/to/TNSStudio/app/modules/recorder/recorder.module.ts`.
4. ERROR in Entry module not found: Error: Can't resolve `./app.css` in `/path/to/TNSStudio/app`.
5. ERROR in [copy-webpack-plugin] unable to locate `app.css` at `/path/to/TNSStudio/app/app.css`.

The first three errors are purely Angular **Ahead of Time** (**AoT**) compilation related. The last two are purely related to webpack configuration. Let's look at each error and how to properly resolve it.

Solution #1: Unexpected value 'SlimSliderDirective...'

Consider the first complete error mentioned in the preceding section:

```
ERROR in Unexpected value 'SlimSliderDirective in
/path/to/TNSStudio/app/modules/player/directives/slider.directive.d.ts'
declared by the module 'PlayerModule in
/path/to/TNSStudio/app/modules/player/player.module.ts'. Please add a
@Pipe/@Directive/@Component annotation.
```

The solution to the preceding error is to install an additional webpack plugin:

```
npm install nativescript-webpack-import-replace --save-dev
```

Then, open `webpack.config.js` and configure the plugin as follows:

```
function getPlugins(platform, env) {
    let plugins = [
      ...
      new ImportReplacePlugin({
          platform: platform,
          files: [
              'slider.directive'
          ]
      }),
      ...
```

This will find the `slider.directive` import in `app/modules/players/directives/index.ts` and append the correct target platform suffix, so the AoT compiler will pick up the right target platform implementation file.

At the time of writing this book, a solution did not exist for that error, so we developed the `nativescript-webpack-import-replace` plugin to solve. Since you might encounter situations with webpack bundling that may require some additional webpack help via a plugin, we will share an overview of how we developed the plugin to solve that error in the event that you encounter other obscure errors that might require you to create a plugin.

Let's look at solutions for the initial remaining errors first, and then we'll highlight webpack plugin development.

Solution #2: Cannot determine the module for class SlimSliderDirective...

Consider the second complete error mentioned in the *Preparing to use webpack* section:

```
ERROR in Cannot determine the module for class SlimSliderDirective in
/path/to/TNSStudio/app/modules/player/directives/slider.directive.android.t
s! Add SlimSliderDirective to the NgModule to fix it.
Cannot determine the module for class SlimSliderDirective in
/path/to/TNSStudio/app/modules/player/directives/slider.directive.ios.ts!
Add SlimSliderDirective to the NgModule to fix it.
```

The solution to the preceding error is to open `tsconfig.aot.json`, and make the following change:

```
BEFORE:
  ...
  "exclude": [
    "node_modules",
    "platforms"
  ],

AFTER:
  ...
  "files": [
    "./app/main.ts"
  ]
```

Since AoT compilation uses the `tsconfig.aot.json` configuration, we want to be more specific with the files that are targeted for compilation. Since `./app/main.ts` is our entry point to bootstrap the app, we will target that file and remove the `exclude` block.

If we were to try bundling now at this point, we would have solved the error we saw; however, we would see the following *new* errors:

```
ERROR in .. lazy
Module not found: Error: Can't resolve
'/path/to/TNSStudio/app/modules/mixer/mixer.module.ngfactory.ts' in
'/path/to/TNSStudio'
 @ .. lazy
 @ ../~/@angular/core/@angular/core.es5.js
 @ ./vendor.ts
```

```
ERROR in .. lazy
Module not found: Error: Can't resolve
'/path/to/TNSStudio/app/modules/recorder/recorder.module.ngfactory.ts' in
'/path/to/TNSStudio'
 @ .. lazy
 @ ../~/@angular/core/@angular/core.es5.js
 @ ./vendor.ts
```

This is because we are targeting our `./app/main.ts`, which branches out to all other imports to our app's files, except to those modules that are lazy loaded.

The solution to the preceding error is to add lazy-loaded module paths in the `files` section:

```
"files": [
  "./app/main.ts",
  "./app/modules/mixer/mixer.module.ts",
  "./app/modules/recorder/recorder.module.ts"
 ],
```

Okay, we solved the `lazy` error; however, now this reveals several *new* errors, as follows:

```
ERROR in
/path/to/TNSStudio/app/modules/recorder/components/record.component.ts
(128,19): Cannot find name 'CFRunLoopGetMain'.
ERROR in
/path/to/TNSStudio/app/modules/recorder/components/record.component.ts
(130,9): Cannot find name 'CFRunLoopPerformBlock'.
ERROR in
/path/to/TNSStudio/app/modules/recorder/components/record.component.ts
(130,40): Cannot find name 'kCFRunLoopDefaultMode'.
ERROR in
/path/to/TNSStudio/app/modules/recorder/components/record.component.ts
(131,9): Cannot find name 'CFRunLoopWakeUp'.
```

Right about now...

The funk soul brother.

Yes, you might be singing Fatboy Slim or about to lose your mind, and we understand. Bundling with webpack can be quite an adventure at times. The best advice that we can provide is to maintain patience and diligence to handle one error at a time; we're almost there.

The solution to the preceding error is to include the iOS and Android platform declarations since we are using native APIs in our app:

```
"files": [
  "./app/main.ts",
  "./app/modules/mixer/mixer.module.ts",
  "./app/modules/recorder/recorder.module.ts",
  "./node_modules/tns-platform-declarations/ios.d.ts",
  "./node_modules/tns-platform-declarations/android.d.ts"
]
```

Hooray, we have now fully resolved the second issue. Let's move on to the next one.

Solution #3: Error encountered resolving symbol values statically

Consider the third complete error mentioned in the *Preparing to use webpack* section:

```
ERROR in Error encountered resolving symbol values statically. Calling
function 'ModalDialogParams', function calls are not supported. Consider
replacing the function or lambda with a reference to an exported function,
resolving symbol RecorderModule in
/path/to/TNSStudio/app/modules/recorder/recorder.module.ts, resolving
symbol RecorderModule in
/path/to/TNSStudio/app/modules/recorder/recorder.module.ts
```

The solution to the preceding error is to open `app/modules/recorder/recorder.module.ts` and make the following change:

```
...
// factory functions
export function defaultModalParamsFactory() {
  return new ModalDialogParams({}, null);
};
...
@NgModule({
  ...
  providers: [
    ...PROVIDERS,
    {
      provide: ModalDialogParams,
      useFactory: defaultModalParamsFactory
    }
  ],
  ...
})
```

```
export class RecorderModule { }
```

This will satisfy the Angular AoT compiler's need to resolve symbols statically.

Solution #4 and #5: Can't resolve './app.css'

Consider the 4th and 5th errors mentioned in the *Preparing to use webpack* section:

```
4. ERROR in Entry module not found: Error: Can't resolve './app.css' in
'/path/to/TNSStudio/app'

5. ERROR in [copy-webpack-plugin] unable to locate 'app.css' at
'/path/to/TNSStudio/app/app.css'
```

The solution to the preceding error is actually related to the fact that we are using platform-specific `.ios.css` and `.android.css`, which is compiled via SASS. We need to update our webpack config so that it knows this. Open `webpack.config.js`, which the plugin added for us automatically, and make the following changes:

```
module.exports = env => {
  const platform = getPlatform(env);

  // Default destination inside platforms/<platform>/...
  const path = resolve(nsWebpack.getAppPath(platform));

  const entry = {
    // Discover entry module from package.json
    bundle: `./${nsWebpack.getEntryModule()}`,
    // Vendor entry with third-party libraries
    vendor: `./vendor`,
    // Entry for stylesheet with global application styles
    [mainSheet]: `./app.${platform}.css`,
  };
  ...

 function getPlugins(platform, env) {
  ...
  // Copy assets to out dir. Add your own globs as needed.
  new CopyWebpackPlugin([
    { from: "app." + platform + ".css", to: mainSheet },
    { from: "css/**" },
    { from: "fonts/**" },
    { from: "**/*.jpg" },
    { from: "**/*.png" },
    { from: "**/*.xml" },
  ], { ignore: ["App_Resources/**"] }),
```

```
...
```

Okay, we now have the bundling issues all cleared, or wait....**do we?!**

We haven't tried to run the app yet in a simulator or on a device. If we were to try and do this now using `npm run start-ios-bundle` or via XCode or `npm run start-android-bundle`, you might get an app crash right when it tries to boot with an error like this:

```
JS ERROR Error: No NgModule metadata found for 'AppModule'.
```

The solution to the preceding error is to ensure that your app contains an `./app/main.aot.ts` file with the following contents:

```
import { platformNativeScript } from "nativescript-angular/platform-static";
import { AppModuleNgFactory } from "./app.module.ngfactory";

platformNativeScript().bootstrapModuleFactory(AppModuleNgFactory);
```

If you recall we have a demo composition setup which loads it's track files from an `audio` folder. We also utilize font-awesome icons with the help of a font-awesome.css file loaded from an `assets` folder. We need to make sure these folders also get copied into our production webpack build. Open `webpack.config.js` and make the following change:

```
new CopyWebpackPlugin([
  { from: "app." + platform + ".css", to: mainSheet },
  { from: "assets/**" },
  { from: "audio/**" },
  { from: "css/**" },
  { from: "fonts/**" },
  { from: "**/*.jpg" },
  { from: "**/*.png" },
  { from: "**/*.xml" },
], { ignore: ["App_Resources/**"] }),
```

SUCCESS!

We can now run our bundled app with no errors using the following commands:

- `npm run start-ios-bundle`
- Opening the XCode project and running `npm run start-android-bundle`

It's worth noting that all the changes we made to enable webpack bundling for release of our app also work perfectly well in development, so rest assured that you have only improved your app's setup at this point.

Detour – Overview of developing a webpack plugin

We now want to return to the first error we encountered when bundling our app which was:

- ERROR in Unexpected value `SlimSliderDirective` in `/path/to/TNSStudio/app/modules/player/directives/slider.directive.d.ts` declared by the module `PlayerModule` in `/path/to/TNSStudio/app/modules/player/player.module.ts`. Please add a `@Pipe/@Directive/@Component` annotation.

A solution for this error did not exist at the time of writing this book, so we created the `nativescript-webpack-import-replace` (`https://github.com/NathanWalker/nativescript-webpack-import-replace`) plugin to solve the problem.

Developing a webpack plugin in detail is out of the scope of this book, but we wanted to give you some highlights to the process in case you end up needing to create one to solve a particular case for your app.

We started by creating a separate project with a `package.json` file so we could install our webpack plugin like any other npm plugin:

```
{
  "name": "nativescript-webpack-import-replace",
  "version": "1.0.0",
  "description": "Replace imports with .ios or .android suffix for target
mobile platforms.",
  "files": [
    "index.js",
    "lib"
  ],
  "engines": {
    "node": ">= 4.3 < 5.0.0 || >= 5.10"
  },
  "author": {
    "name": "Nathan Walker",
    "url": "http://github.com/NathanWalker"
  },
  "keywords": [
    "webpack",
    "nativescript",
    "angular"
```

```
  ],
  "nativescript": {
    "platforms": {
      "android": "3.0.0",
      "ios": "3.0.0"
    },
    "plugin": {
      "nan": "false",
      "pan": "false",
      "core3": "true",
      "webpack": "true",
      "category": "Developer"
    }
  },
  "homepage":
"https://github.com/NathanWalker/nativescript-webpack-import-replace",
  "repository": "NathanWalker/nativescript-webpack-import-replace",
  "license": "MIT"
}
```

The `nativescript` key actually helps categorize this plugin on the various NativeScript plugin listing sites.

We then created `lib/ImportReplacePlugin.js` to represent the actual plugin class we would be able to import and use in our webpack config. We created this file inside a `lib` folder for good measure in case we need to add extra supporting files to aid our plugin for a nice clean separation of concerns with our plugin organization. In this file, we set up an export by defining a closure containing a constructor for our plugin:

```
exports.ImportReplacePlugin = (function () {
  function ImportReplacePlugin(options) {
    if (!options || !options.platform) {
      throw new Error(`Target platform must be specified!`);
    }

    this.platform = options.platform;
    this.files = options.files;
    if (!this.files) {
      throw new Error(`An array of files containing just the filenames to
replace with platform specific names must be specified.`);
    }
  }

  return ImportReplacePlugin;
})();
```

This will take the target `platform` defined in our webpack config and pass it through as options along with a `files` collection, which will contain all the filenames of the imports we need to replace.

We then want to wire into webpack's `make` lifecycle hook to grab hold of the source files being processed in order to parse them:

```
ImportReplacePlugin.prototype.apply = function (compiler) {
    compiler.plugin("make", (compilation, callback) => {
      const aotPlugin = getAotPlugin(compilation);
      aotPlugin._program.getSourceFiles()
        .forEach(sf => {
          this.usePlatformUrl(sf)
        });

      callback();
    })
 };

  function getAotPlugin(compilation) {
    let maybeAotPlugin = compilation._ngToolsWebpackPluginInstance;
    if (!maybeAotPlugin) {
      throw new Error(`This plugin must be used with the AotPlugin!`);
    }
    return maybeAotPlugin;
  }
```

This grabs hold of all the AoT source files. Then we setup a loop to process them one by one and add processing methods for what we need:

```
ImportReplacePlugin.prototype.usePlatformUrl = function (sourceFile) {
    this.setCurrentDirectory(sourceFile);
    forEachChild(sourceFile, node => this.replaceImport(node));
}

ImportReplacePlugin.prototype.setCurrentDirectory = function (sourceFile) {
   this.currentDirectory = resolve(sourceFile.path, "..");
}

ImportReplacePlugin.prototype.replaceImport = function (node) {
    if (node.moduleSpecifier) {
      var sourceFile = this.getSourceFileOfNode(node);
      const sourceFileText = sourceFile.text;
      const result = this.checkMatch(sourceFileText);
      if (result.index > -1) {
        var platformSuffix = "." + this.platform;
        var additionLength = platformSuffix.length;
```

```
        var escapeAndEnding = 2; // usually "\";" or "\';"
        var remainingStartIndex = result.index + (result.match.length - 1)
+ (platformSuffix.length - 1) - escapeAndEnding;

        sourceFile.text =
          sourceFileText.substring(0, result.index) +
          result.match +
          platformSuffix +
          sourceFileText.substring(remainingStartIndex);

        node.moduleSpecifier.end += additionLength;
      }
    }
  }

  ImportReplacePlugin.prototype.getSourceFileOfNode = function (node) {
    while (node && node.kind !== SyntaxKind.SourceFile) {
      node = node.parent;
    }
    return node;
  }

  ImportReplacePlugin.prototype.checkMatch = function (text) {
    let match = '';
    let index = -1;
    this.files.forEach(name => {
      const matchIndex = text.indexOf(name);
      if (matchIndex > -1) {
        match = name;
        index = matchIndex;
      }
    });
    return { match, index };
  }
```

An interesting part to building webpack plugins (*and arguably the most challenging*) is working with **Abstract Syntax Trees** (**ASTs**) of your source code. A critical aspect of our plugin is getting the "source file" node from the AST as follows:

```
ImportReplacePlugin.prototype.getSourceFileOfNode = function (node) {
  while (node && node.kind !== SyntaxKind.SourceFile) {
    node = node.parent;
  }
  return node;
}
```

This effectively weeds out any other nodes that are not source files since that is all our plugin needs to deal with.

Lastly, we create an `index.js` file in the root to simply export the plugin file for use:

```
module.exports = require("./lib/ImportReplacePlugin").ImportReplacePlugin;
```

With the aid of this webpack plugin, we are able to completely solve all the webpack bundling errors we encountered in our app.

Summary

In this chapter, we prepared our app for distribution by adding webpack into our build chain to help ensure that we have the smallest JavaScript size and the optimal execution performance of our code. This also enabled Angular's AoT compilation on our app, which helps to provide an optimal performance of our code.

Along the way, we provided a cookbook of solutions to various webpack bundling errors that you might run into during the course of your app's development. In addition, we took a high-level look at developing a custom webpack plugin to help solve a particular error condition in our app to achieve a successful bundle.

Now that we have an optimal bundle of our app code, we are now ready to finish our distribution steps to finally deploy our app in the next chapter.

15
Deploying to the Apple App Store

In this chapter, we will focus on how to deploy our app to the Apple App Store. There are several important steps we will want to follow, so pay close attention to all the details presented here.

Whether you need to work with Signing Certificates to build a release target of our app, generate app icons and splash screens, or work within XCode to archive our app for upload to the App Store, we will cover all these topics in this chapter.

TJ VanToll, a NativeScript expert and developer advocate for Progress, wrote an excellent article detailing deployment steps titled *8 Steps to Publish Your NativeScript App to the App Stores* (`https://www.nativescript.org/blog/steps-to-publish-your-nativescript-app-to-the-app-stores`). We will take excerpts from that article and expand on sections wherever possible here in this chapter and the next chapter.

> *There's no point in lying to you—releasing an iOS app to the iOS App Store is one of the most painful processes you'll go through in your software development career. So, in case you get stuck or confused in these steps, just know that it's not just you—everyone gets frustrated when releasing iOS apps the first time.*

The following topics are covered in this chapter:

- How to create an App ID and production certificate to sign your app release target with
- How to configure a NativeScript app with the appropriate metadata needed for a release
- How to handle app icons and splash screens
- Uploading your build to iTunes Connect using the NativeScript CLI

Preparing for App Store distribution

To deploy iOS apps to the iOS App Store, you absolutely must have an active Apple Developer account. It costs $99 USD per year to be a part of the program, and you can sign up at `developer.apple.com/register`.

App ID, certificates, and profiles

Once you create an Apple Developer account, you'll need to create an App ID, Production Certificate, and a Distribution Provisioning Profile on the Apple Developer portal. This is the most tedious part of the entire process, as it takes some time to learn what each of these various files do and how to use them:

1. For our app, we will begin by creating the App ID with the following:

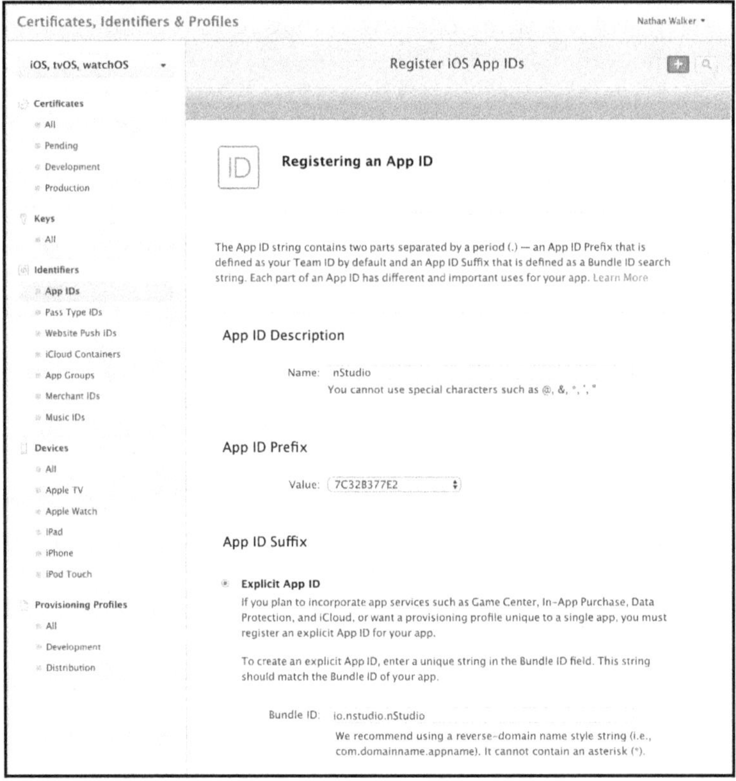

2. Once we create this App ID, we can now create a **Production** certificate:

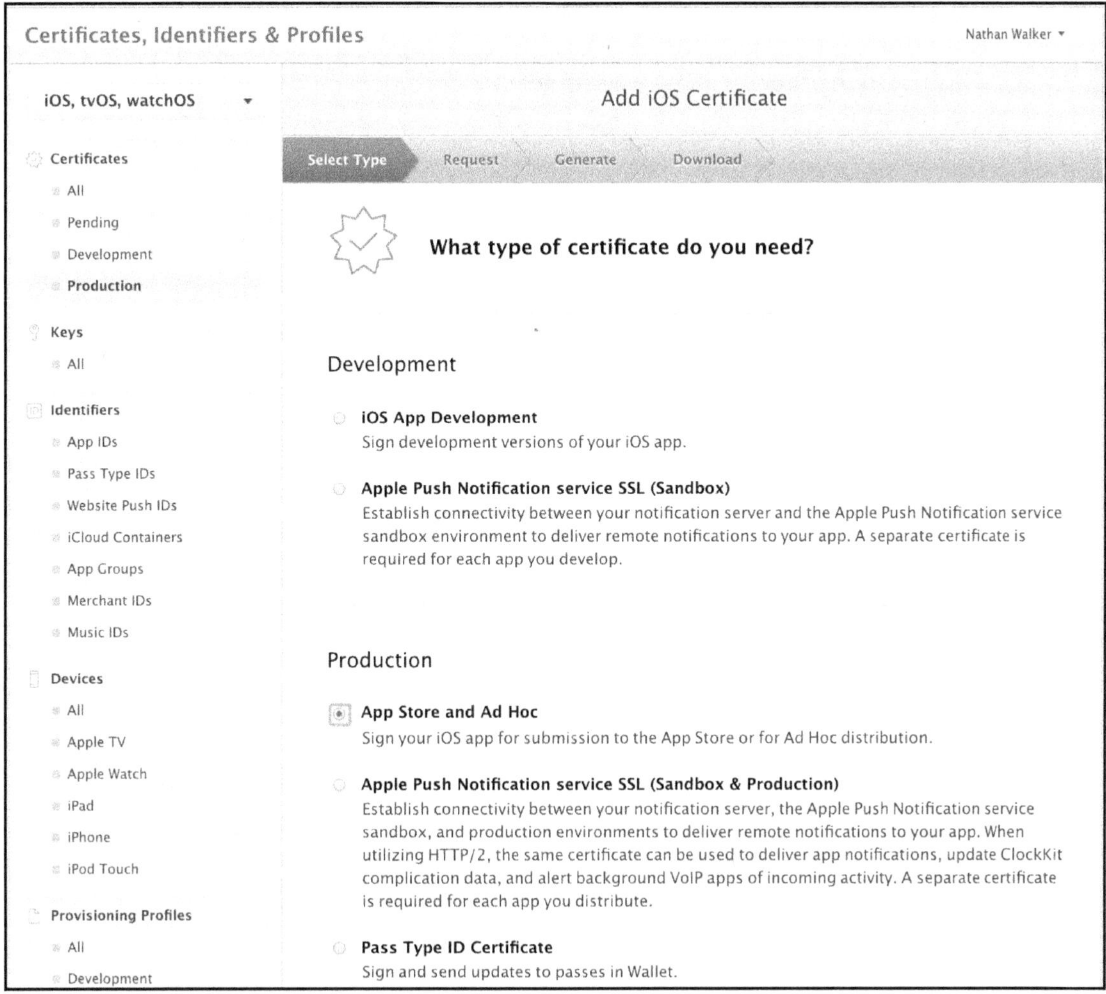

3. Select **Continue**. Then, the next screen will provide instructions on how to sign your production certificate, which we will walk through next. First, open `/Applications/Utilities/Keychain Access.app` and then go to the top-left menu and select **Certificate Assistant** | **Request a Certificate from a Certificate Authority** using a setup like this:

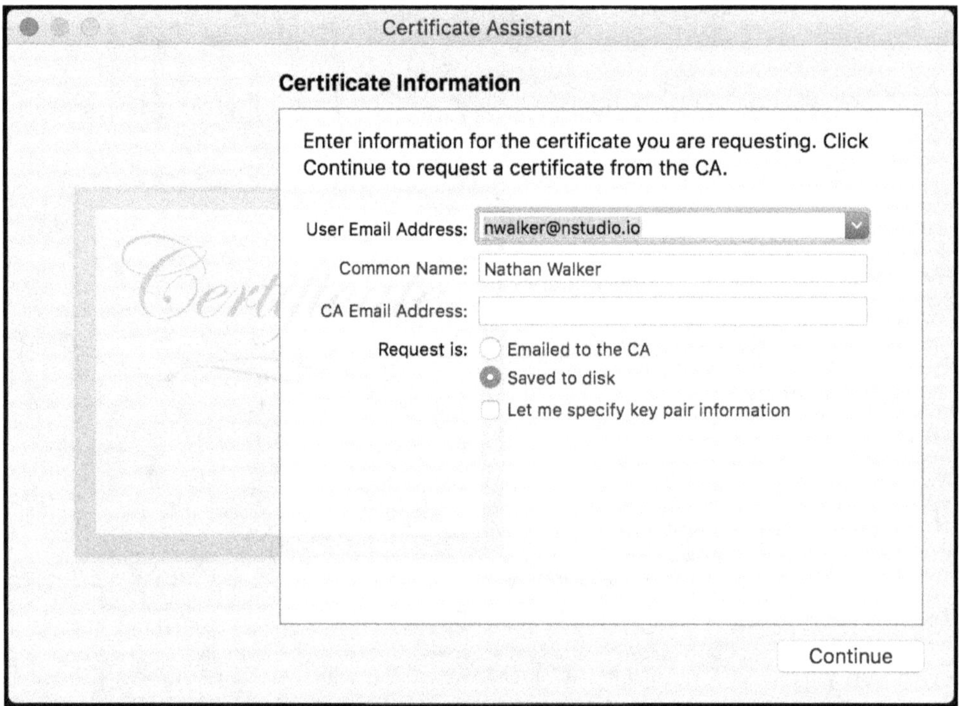

This will save a signing request file wherever you choose, which you will need in the next step.

4. Now, select that signing request file at this step in the portal:

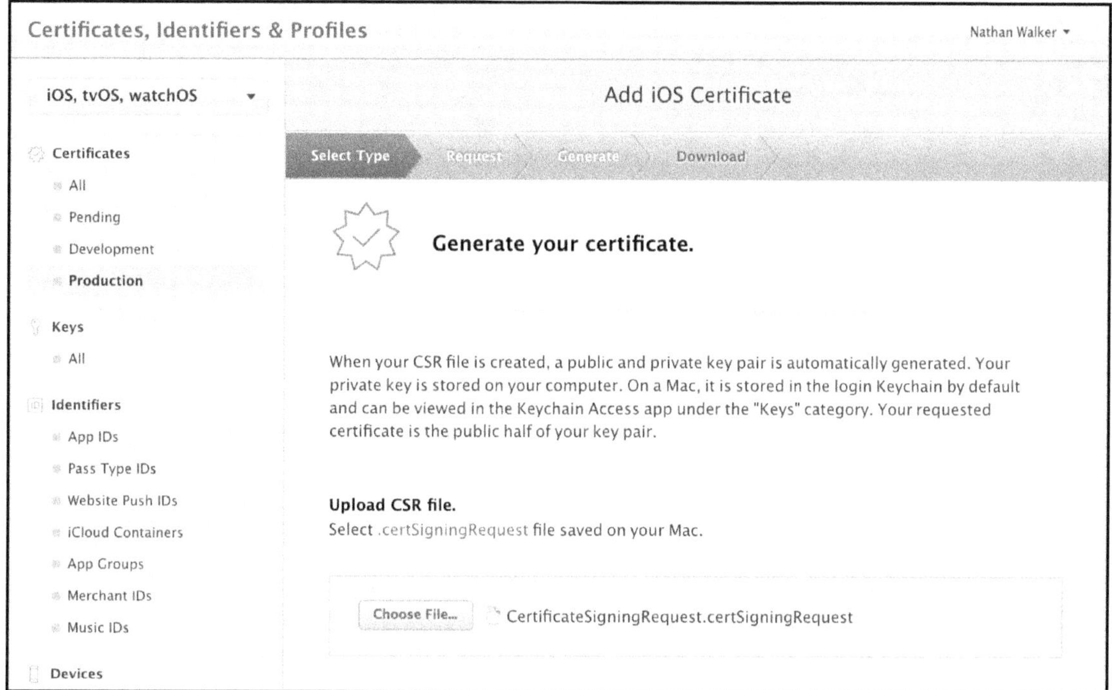

5. On the next screen, it's very important to download and then double-click the file that you need to install into your keychain as it specifies:

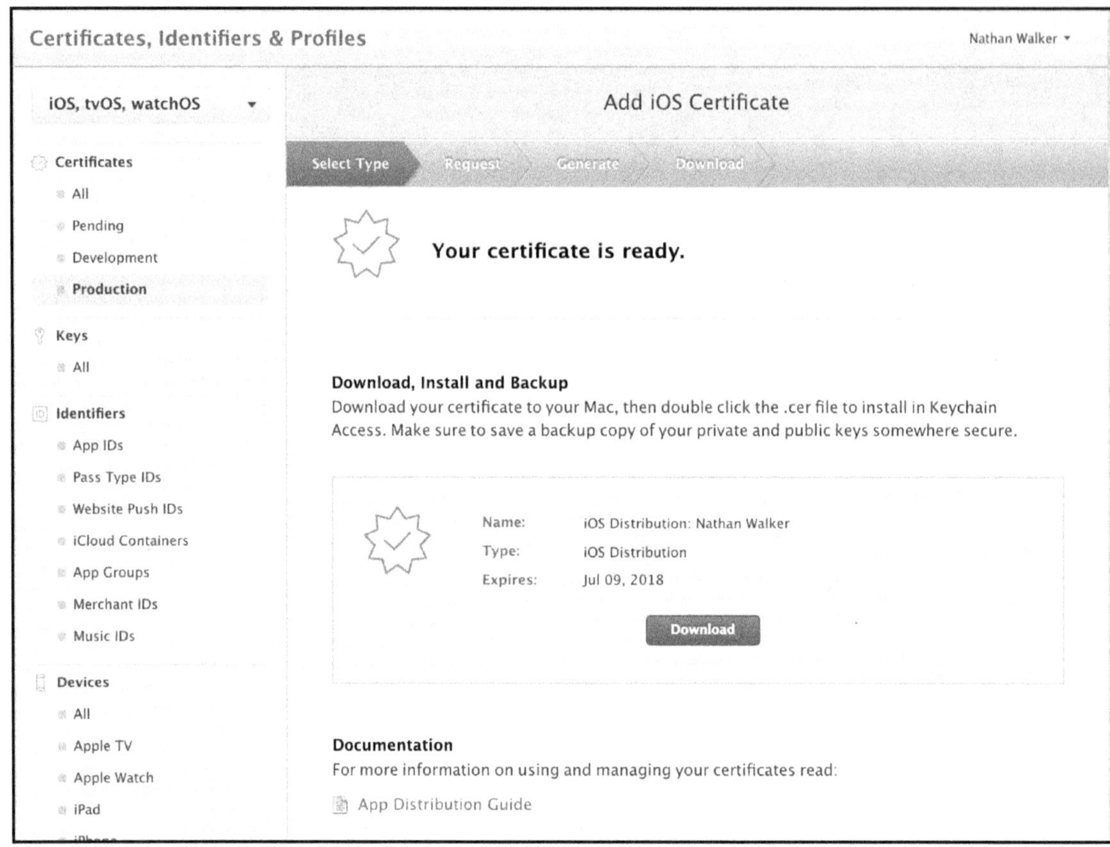

6. When double-clicking the file to install into the keychain, it may prompt you to provide the keychain to install the file into; using the *login* keychain will work fine:

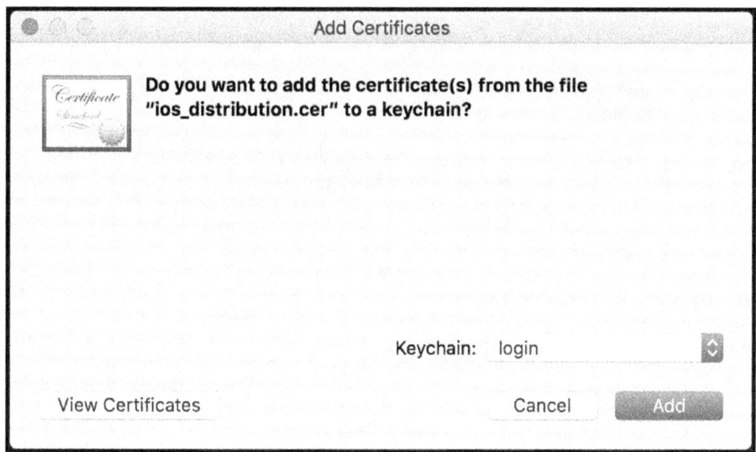

You should now see something similar to the following screenshot in your keychain access app:

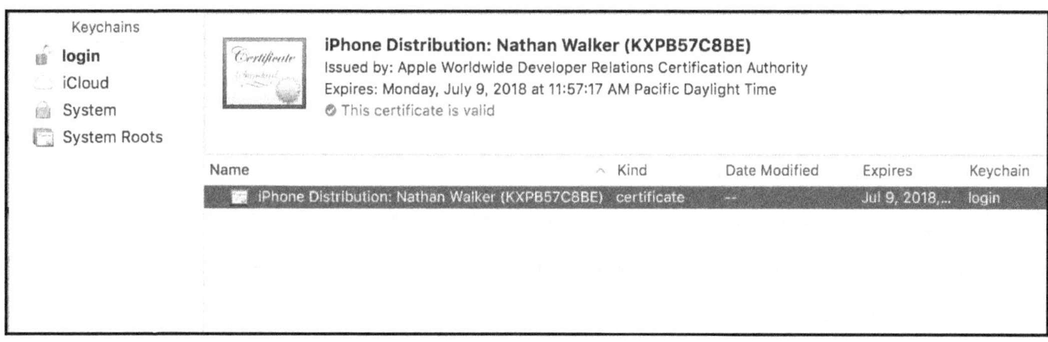

7. You can now quit keychain access.

8. Next, we want to create a Distribution Provisioning Profile:

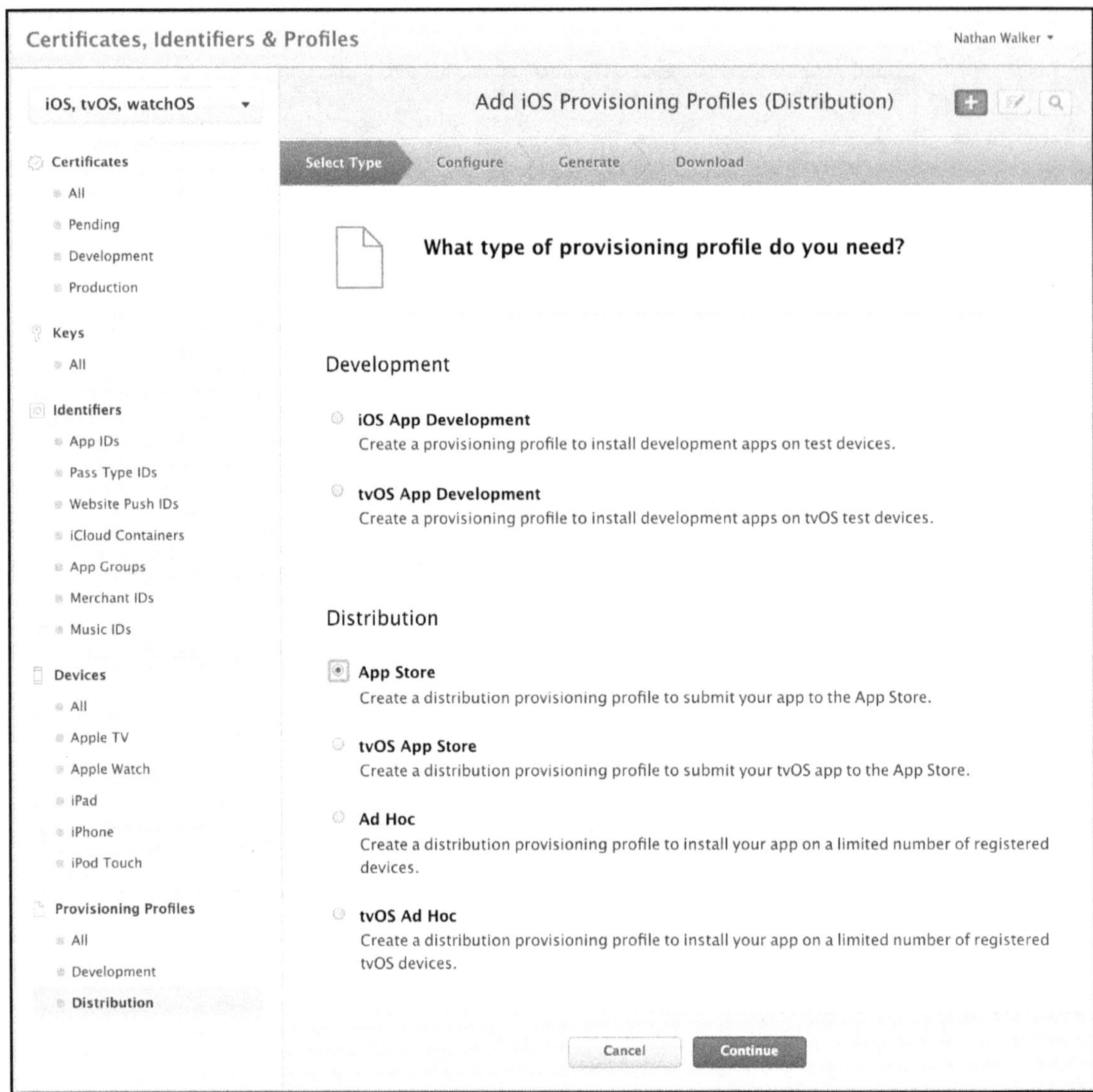

9. On the next screen, just ensure that you select the App ID that you created:

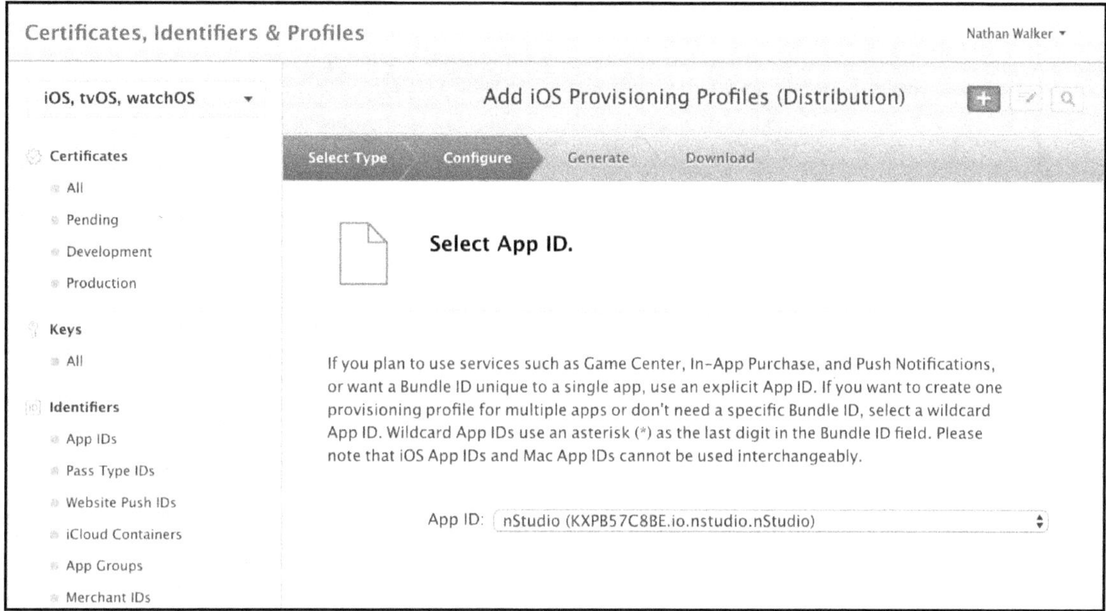

10. Then, on the next screen, you should be able to select the **Distribution** certificate you created:

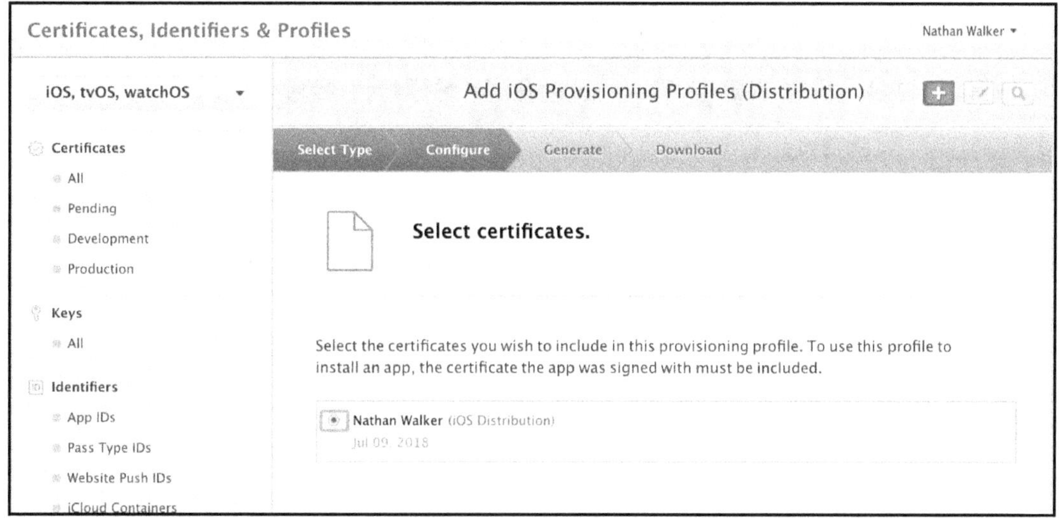

11. Then, you will be able to give the profile a name:

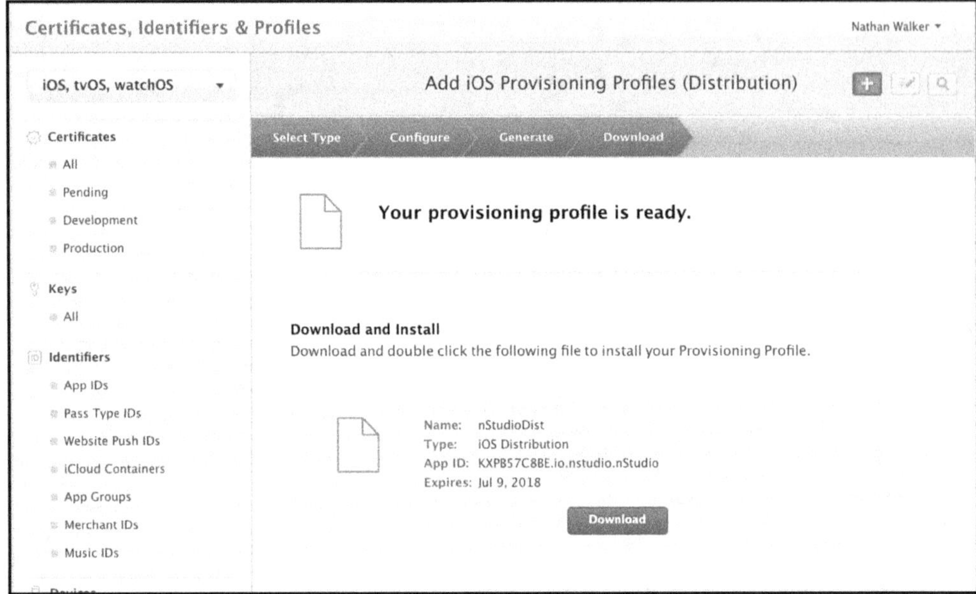

12. You can download the profile and just place it alongside your `ios_distribution.cer` file; however, there is no need to open that provisioning profile file, as XCode will handle everything else.

Configure the app metadata such as App ID and Display Name

iOS and Android apps have a lot of information that you will need to configure before you deploy your apps to their respective stores. NativeScript provides intelligent defaults for many of these values, but there are a few that you'll want to review before you deploy.

App ID

The App ID configured in the Apple developer portal moments ago is a unique identifier for your app that uses something called reverse domain name notation. Our NativeScript app's metadata must match. Our App ID for this app is `io.nstudio.nStudio`. The NativeScript CLI has a convention for setting the App ID during app creation:

```
tns create YourApp --appid com.mycompany.myappname
```

We did not use this option when we created our app; however, it's easy enough to change our App ID.

Open the app's root `package.json` file and find the `nativescript` key. Ensure that the `id` attribute contains the value you'd like to use:

```
"nativescript": {
  "id": "io.nstudio.nStudio",
  "tns-ios": {
    "version": "3.1.0"
  },
  "tns-android": {
    "version": "3.1.1"
  }
},
```

Display name

You app's display name is the name the user sees next to your icon on their screen. By default, NativeScript sets your app's display name based on the value you passed to `tns create`, which is oftentimes not exactly what you want the user to see. For example, running `tns create my-app` results in an app with a display name of `myapp`.

To change that value on iOS, first open your app's `app/App_Resources/iOS/Info.plist` file. The `Info.plist` file is iOS's main configuration file, and here you'll find a number of values you may want to tinker with before releasing your app. For display name, you'll want to alter the `CFBundleDisplayName` value.

Here's what this value looks like for `nStudio`:

```
<?xml version="1.0" encoding="UTF-8"?>
<!DOCTYPE plist PUBLIC "-//Apple//DTD PLIST 1.0//EN"
"http://www.apple.com/DTDs/PropertyList-1.0.dtd">
<plist version="1.0">
<dict>
  <key>CFBundleDevelopmentRegion</key>
  <string>en</string>
  <key>CFBundleDisplayName</key>
  <string>nStudio</string>
  <key>CFBundleExecutable</key>
  <string>${EXECUTABLE_NAME}</string>
  <key>CFBundleInfoDictionaryVersion</key>
  <string>6.0</string>
```

Although there's no real character limit to display names, both iOS and Android will truncate your display names after somewhere around 10–12 characters.

Create your app icons and splash screens

Your app's icon is the first thing your users notice about your app. When you start a new NativeScript app, you will get a placeholder icon, which is fine for development; however, for production, you will need to replace the placeholder icon with the image you'll want to go to the stores with.

To get your production-ready app icon files in place, you will need to first create a 1024 x 1024 pixel `.png` image asset that represents your app.

To make your life difficult, both iOS and Android require you to provide a variety of icon images in a wide array of sizes. Don't worry though; once you have a 1024 x 1024 image, there are a few sites that will generate images in the various dimensions that Android and iOS require. For the NativeScript development, I recommend that you use the Nathanael Anderson's NativeScript Image Builder, which is available at `images.nativescript.rocks`.

We will build our icon in Photoshop:

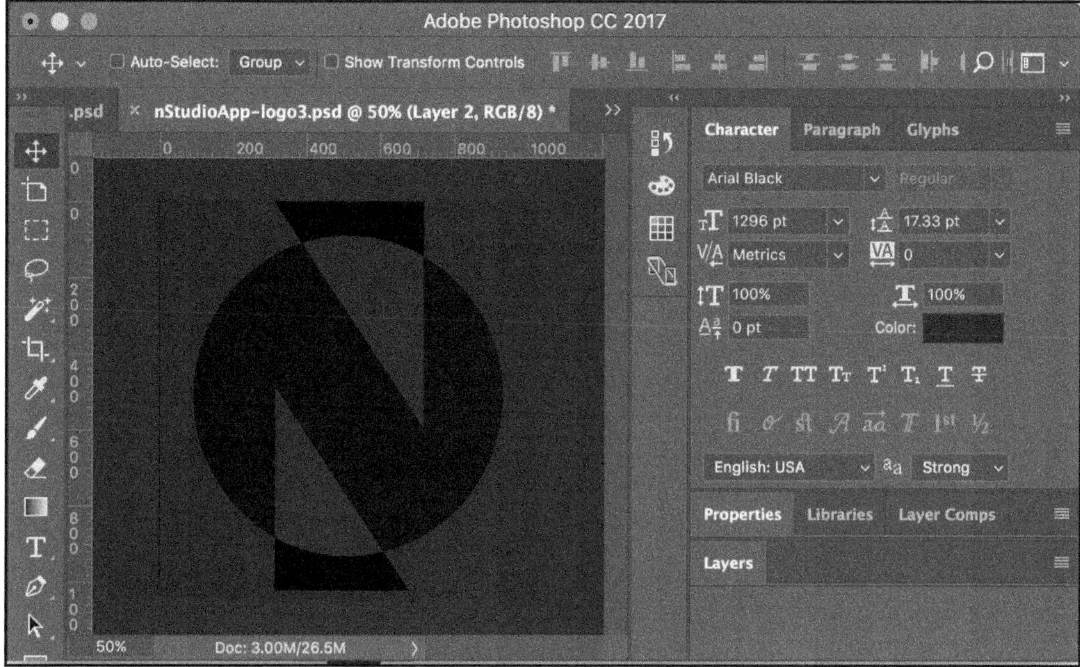

We can then export that as a `.png` and upload it to `images.nativescript.rocks`:

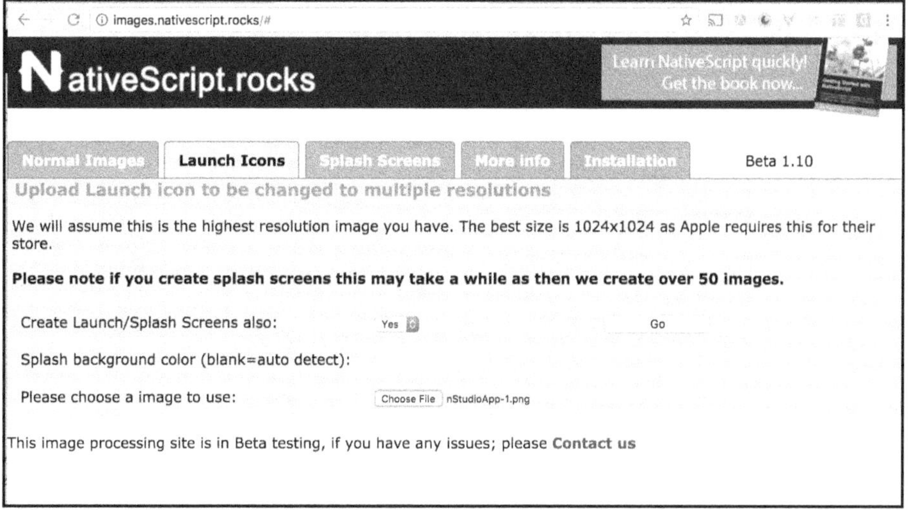

When you click on **Go**, a zip file will be downloaded, and the contents will include your app icons and splash screens. You can copy these images into your `app/App_Resources` folder, respectively, for iOS (we will cover Android in the next chapter).

We now have our app icon and splash screens in place.

Build the app for release

Since we have already covered webpack bundling issues in the preceding chapter, we are now ready to build the final releasable bundle with the following command:

```
npm run build-ios-bundle -- --release --forDevice --teamId KXPB57C8BE
```

Note that the `--teamId` will be different for you. It is the prefix on the App ID provided in the preceding command.

After this command finishes, you'll have the `.ipa` file you'll need in your `platforms/ios/build/device` folder. Make a note of the location of that file, as you'll need it in the final step of this guide.

Phew! Hopefully, you've made it to this point in one piece. You're now ready for the final step, iTunes Connect.

Upload to iTunes Connect

The first thing you'll need to do is register your app. To do that, visit `https://itunesconnect.apple.com/`, click on **My Apps**, and, click on the + button (currently in the top-left corner of the screen), and then select **New App**. In the screen that follows, ensure that you select the correct **Bundle ID**, and the **SKU** can be any number you'd like to identify your app; we like to use the current date:

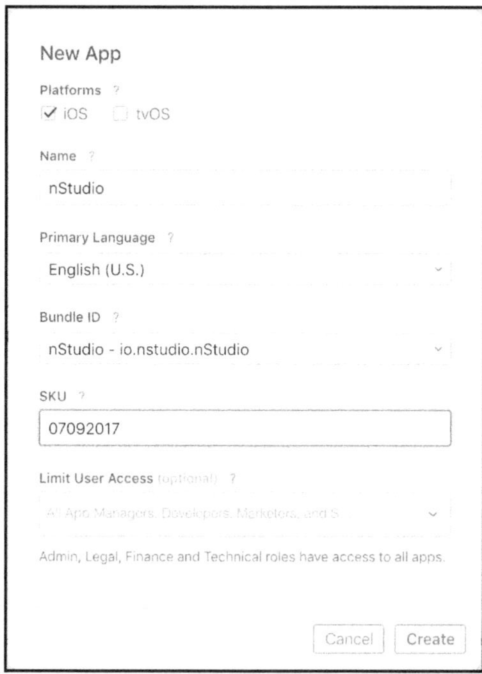

After providing this information, you'll be taken to your app's dashboard where we need to provide more metadata about our app. Most of this information is pretty straightforward, such as descriptions and pricing, but there are a few *fun* pieces we'll have to deal with, such as screenshots.

iTunes Connect now requires you to upload two sets of screenshots, one for the largest iPhone devices (5.5-inch displays), and another for the largest iPad devices (12.9-inch devices). Apple still gives you the ability to provide optimized screenshots for each and every iOS device dimension, but if you provide only 5.5-inch and 12.9-inch screenshots, Apple will rescale your provided screenshots for smaller display devices automatically.

To get those screenshots we could run the app on a physical iPhone Plus and iPad Pro device, but we find it far easier to get these screenshots from iOS simulators.

With the correct simulated device running, we can use the simulator's *Cmd* + *S* keyboard shortcut to take a screenshot of the app, which will save the appropriate image to our desktop.

At this point, we're all set. We will use a service such as DaVinci (`https://www.davinciapps.com`) to polish our image files, but when we are ready, we'll drag our images into the **App Preview and Screenshots** area of iTunes Connect.

Uploading your .ipa file

We're almost there! Once all the information has been entered into iTunes Connect, the final step is to associate the built `.ipa` file with all the information we just typed out.

We will use the NativeScript CLI to do this.

Remember that your .ipa file is in your app's `platforms/ios/build/device` folder.

Run the following command to publish your app to iTunes Connect:

```
tns publish ios --ipa <path to your ipa file>
```

That's it. One important note, though, for whatever crazy reason, there's a nontrivial delay between the time you upload your iOS app and the time that your app shows up in iTunes Connect. We saw that the delay can be as short as 30 seconds and as long as an hour. Once the build shows up there, we can go ahead and click on the big **Submit for Review** button, and cross our fingers.

Apple has a notoriously sporadic delay for reviewing the iOS apps that you submit. At the time of writing this book, the average time to review for the iOS App Store was around 2 days.

Summary

In this chapter, we highlighted the critical steps that must be taken to release an app to the Apple App Store, including signing certificates, app id, app icons, and splash screens. The process may seem heavily involved at first, but once you understand the various steps better, it becomes more clear.

We now have an app pending review in the store and are well on our way to making our app available for users around the world.

In the next chapter, let's finish this by deploying our app to the Google Play Store to broaden our audience.

16
Deploying to Google Play

Although deploying an app to Google Play can be slightly simpler when compared to the Apple App Store, there's still a few key steps that we need to pay attention to. We covered some preparation steps in `Chapter 14`, *Deployment Preparation with webpack Bundling*, and `Chapter 15`, *Deploying to the Apple App Store*, such as using webpack to bundle the app and preparing app icons and splash screens, so we will jump right into building a releasable APK.

We express our gratitude to TJ VanToll for an excellent eight-step article to deploy NativeScript apps (`https://www.nativescript.org/blog/steps-to-publish-your-nativescript-app-to-the-app-stores`) from which we will insert excerpts from and expand wherever possible.

The following topics are covered in this chapter:

- Generating a keystore to build your APK with
- Building a releasable APK with the NativeScript CLI
- Uploading the APK to Google Play for publication

Building an APK for Google Play

Before you open Google Play to register and publish this app (which is the next step), let's double-check a couple of things to ensure that our metadata is correct.

Open `app/App_Resources/Android/app.gradle` and ensure that the `applicationId` is correct for your package name:

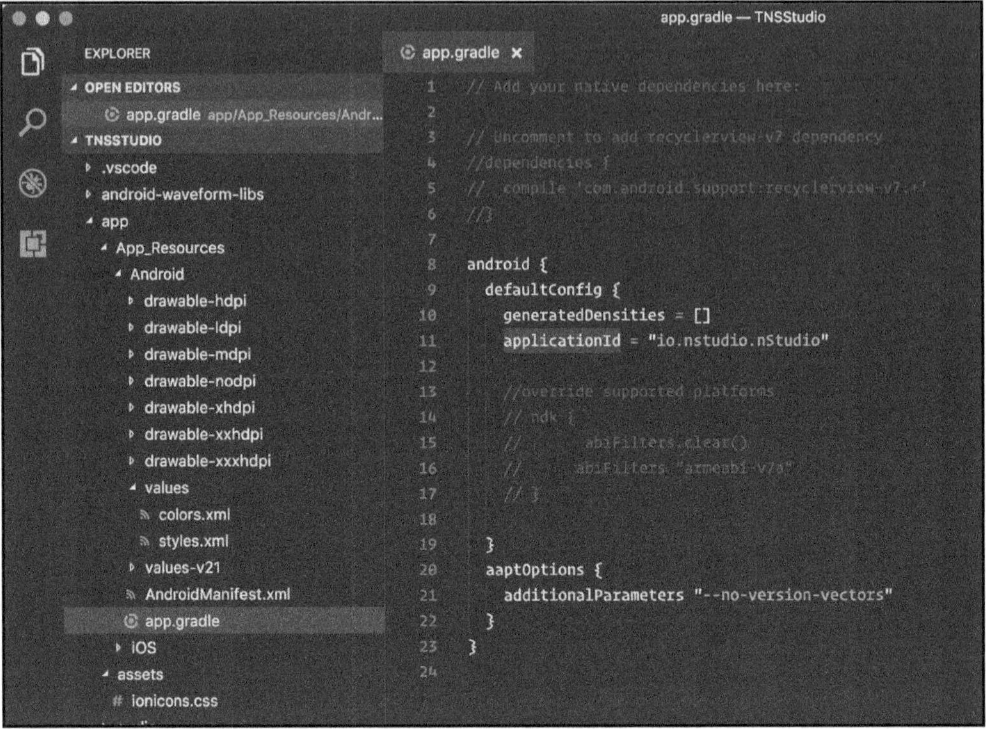

Also, open `package.json` at the root of the project and double-check the `nativescript.id` there as well for good measure:

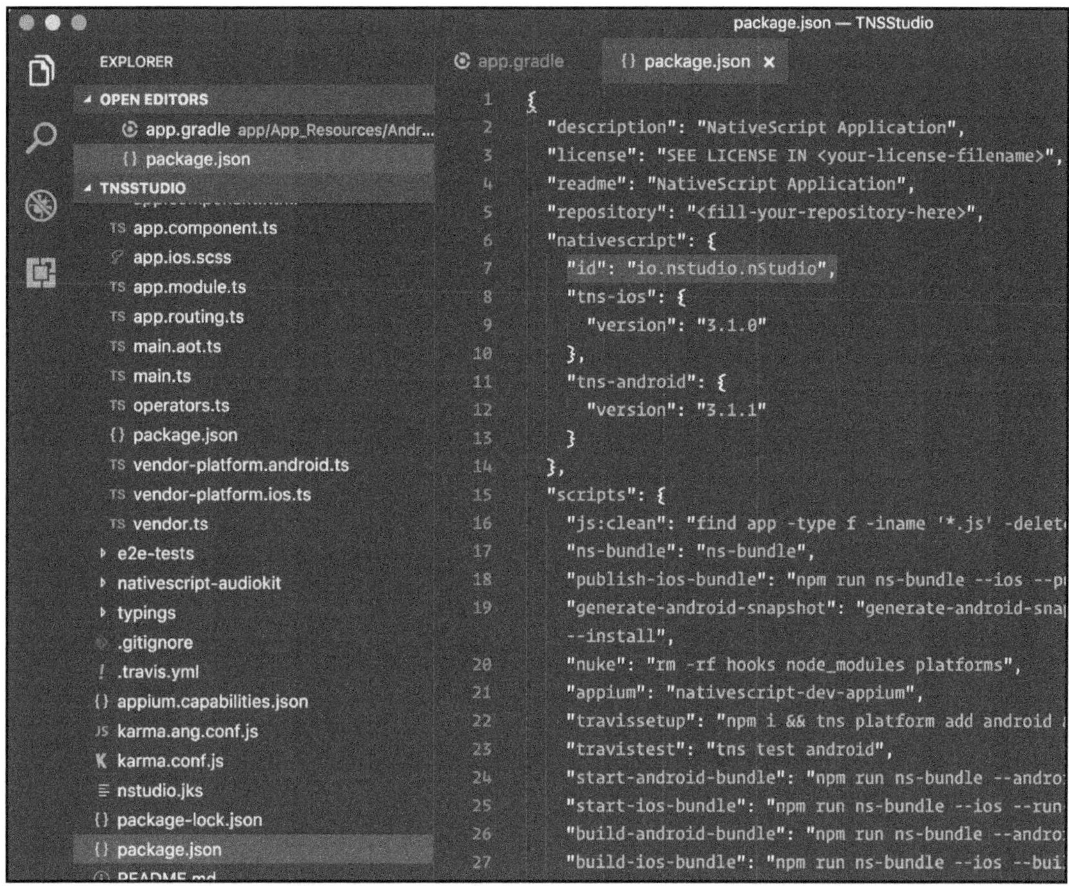

Now, you will need to generate an executable Android file for your application. On Android, this file has a `.apk` extension, and you can generate this file using the NativeScript CLI.

The `tns run` command you were using during NativeScript development actually generates a `.apk` file for you and installs that file on an Android emulator or device. However, for a Google Play release, the build you create must also be code signed. You can refer to Android's documentation (`https://developer.android.com/studio/publish/app-signing.html`) on code signing if you want to dive into the cryptographic details, but at a high level, you will need to do the following two things to create a release version of your Android app:

- Create a `.keystore` or `.jks` (Java keystore) file
- Use that `.keystore` or `.jks` file to sign in to your app during a build

The Android documentation provides you a few options on how you can create your keystore file (`https://developer.android.com/studio/publish/app-signing.html#release-mode`). Our preferred approach is the `keytool` command-line utility, which is included in the Java JDK that NativeScript installs for you, so it should already be available on your development machine's command line.

To use `keytool` to generate a keystore for code signing our app, we will use the following command:

```
keytool -genkey -v -keystore nstudio.jks -keyalg RSA -keysize 2048 -
validity 10000 -alias nstudio
```

The `keytool` utility will ask you a number of questions, several of which are optional (name of organization and the names of your city, state, and country), but the most important ones are the passwords for both the keystore and the alias (more on that later on). Here's what the `keytool` process looks like when we generate the keystore:

```
NathanMac:TNSStudio nathan$ keytool -genkey -v -keystore nstudio.jks -keyalg RSA -keysize 2048 -validity 10000 -alias nstudio
Enter keystore password:
Re-enter new password:
What is your first and last name?
  [Unknown]:  Nathan Walker
What is the name of your organizational unit?
  [Unknown]:  nStudio
What is the name of your organization?
  [Unknown]:
What is the name of your City or Locality?
  [Unknown]:
What is the name of your State or Province?
  [Unknown]:
What is the two-letter country code for this unit?
  [Unknown]:
Is CN=Nathan Walker, OU=nStudio, O=Unknown, L=Unknown, ST=Unknown, C=Unknown correct?
  [no]:  yes

Generating 2,048 bit RSA key pair and self-signed certificate (SHA256withRSA) with a validity of 10,000 days
        for: CN=Nathan Walker, OU=nStudio, O=Unknown, L=Unknown, ST=Unknown, C=Unknown
Enter key password for <nstudio>
        (RETURN if same as keystore password):
[Storing nstudio.jks]
```

Before we move on to how to use this .jks file, there's one important thing you need to know. Place this .jks file somewhere safe, and do not forget the password for the keystore or for the alias. (Personally, I like using the same password for my keystore and my aliases to simplify my life.) Android requires you to use this exact same .jks file to sign in to any and all updates to your app. This means that if you lose this .jks file, or its password, you will not be able to update your Android app. You'll have to create a brand new entry in Google Play, and your existing users will not be able to upgrade—so take care not to lose it!

Oh, and one more thing to note in most cases is that you'll want to use a single keystore file to sign in to all of your personal or company's Android applications. Remember how you had to pass a -alias flag to the keytool utility, and how that alias had its own password? It turns out that one keystore can have many aliases, and you'll want to create one for each Android app that you build.

Okay, so now that you have this .jks file, and you have it stored in a nice and secure location, the rest of the process is quite easy.

Build our Android app using webpack and pass it the information you just used to create the .jks file. For example, the following command is used to create a release build of nStudio:

```
npm run build-android-bundle -- --release --keyStorePath
~/path/to/nstudio.jks --keyStorePassword our-pass --keyStoreAlias nstudio -
-keyStoreAliasPassword our-alias-pass
```

Once the command finishes running, you'll have a releasable .apk file in your app's platforms/android/build/outputs/apk folder; note the location of that file, as you'll need it in the next step—deploying your app on Google Play:

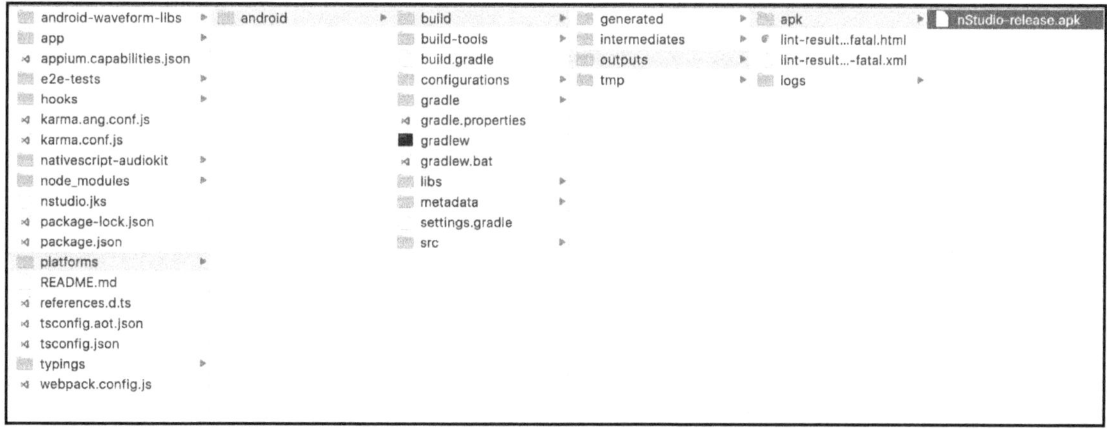

Uploading to Google Play

Google Play is where Android users find and install apps, and the Google Play Developer Console (`https://play.google.com/apps/publish/`) is where developers register and upload apps for users.

You will first create a new application by its name and then see it listed:

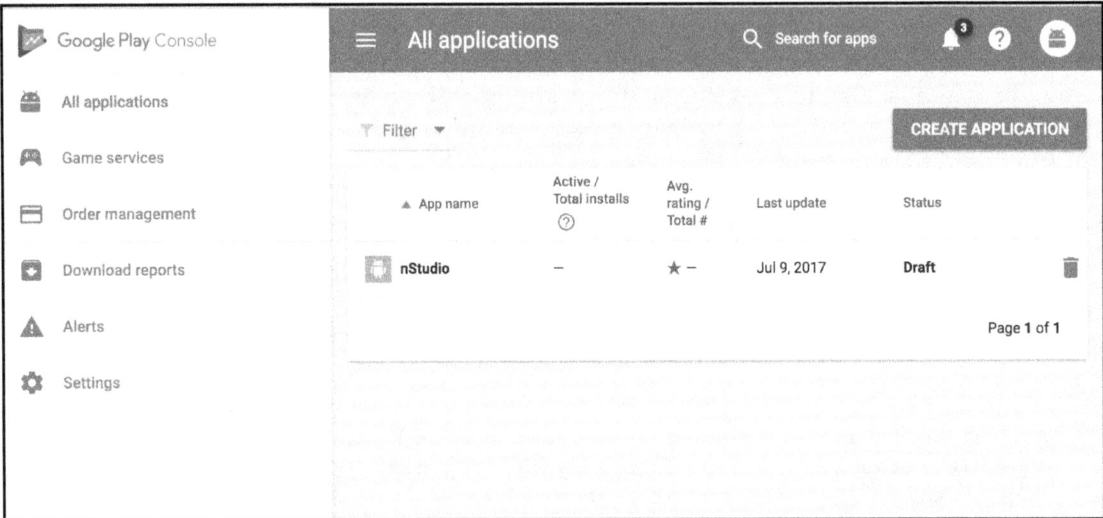

Android's documentation on uploading apps and setting up your store listing is quite good, so we will not recreate all that information here. Instead, a few tips will be provided that you might find helpful when uploading your own NativeScript apps to Google Play.

On the **Store Listing** tab in the Google Play Developer Console, you'll have to provide at least two screenshots of your app in action, as follows:

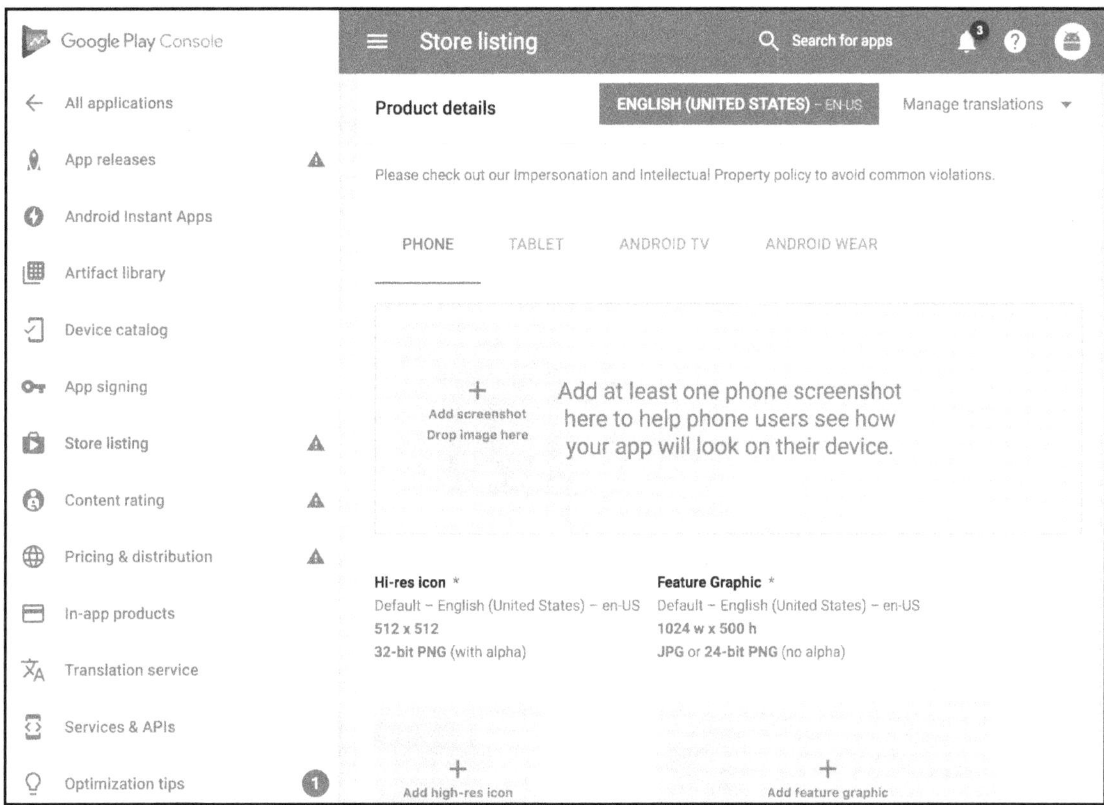

Launch your app in an Android **Android Virtual Device** (**AVD**) using the `tns run android --emulator` command. The Android AVDs have a built-in way of taking screenshots using the little camera icon in the emulator's sidebar.

Use this button to take a few screenshots of the most important screens in your app, and the image files themselves will appear on your desktop. From there, you could take those files and upload them directly into the Google Play Developer Console. A 1024 x 500 **Feature Graphic** image file is also required, which will appear at the top of your store listing, as shown in the following screenshot:

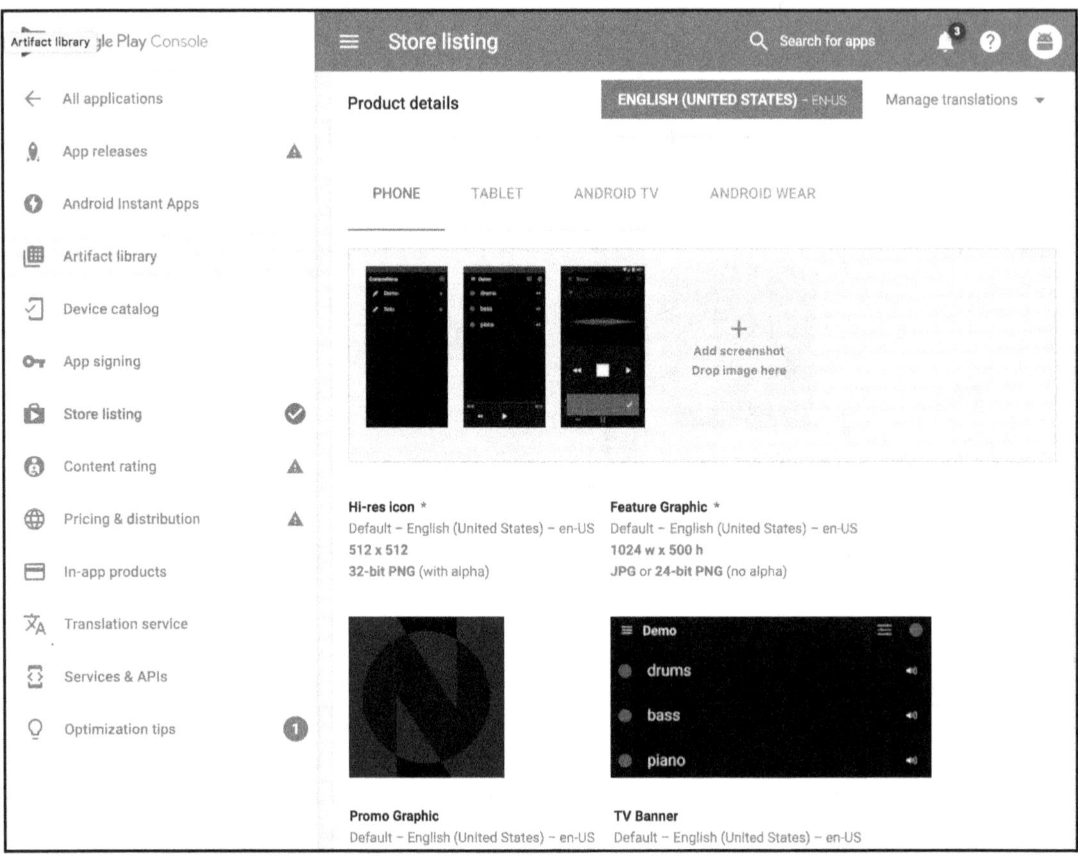

Although not shown in the preceding screenshot, we recommend that you use a service like DaVinci (`https://www.davinciapps.com`) to add a little flair to your screenshots, and turn them into a little tutorial of what your app does.

APK

The **App Releases** section of the Google Play Developer Console is where you upload the `.apk` file you generated in the preceding step of this chapter.

You may see mention of opting into **Google Play App Signing** when you view the **App Releases** section. It's best to opt in now rather than later. Once you've opted in, it will show **Enabled**:

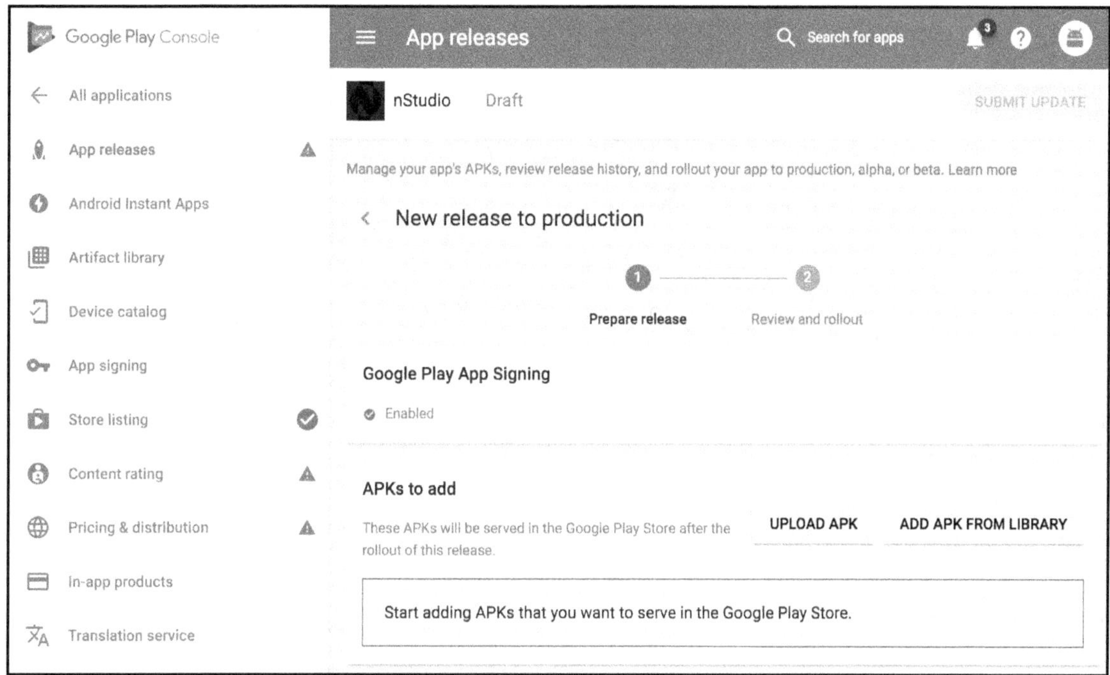

You can then proceed to upload the apk in your app's `platforms/android/build/outputs/apk` folder.

Once you have your APK uploaded, you should see it listed on that same page where you can type out release notes for the uploaded version in multiple languages of your choice:

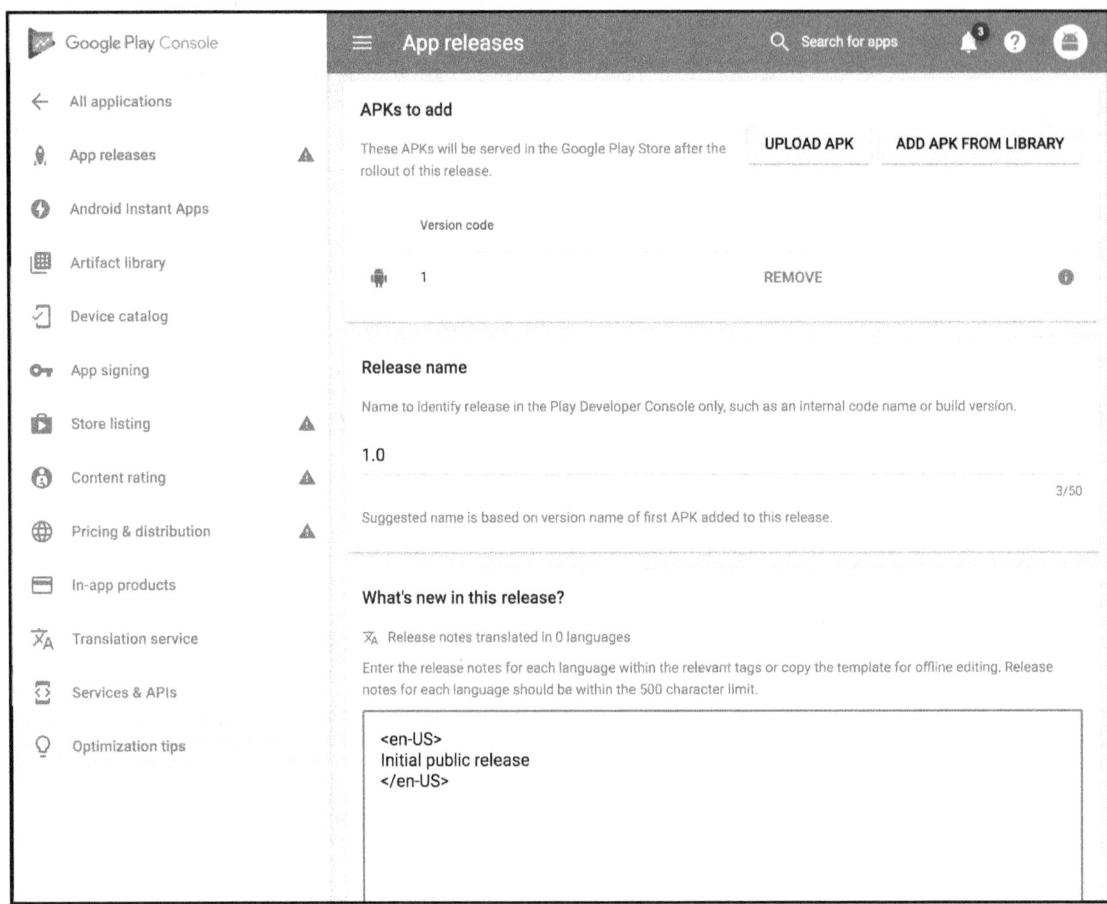

After you click on **Save** on that page, you will probably want to return to the **Store listing** section to finish filling out all of your app's information. Once you have everything set, you're ready to submit your app. Android app reviews generally take a few hours, and unless Google flags any problems, your app should be available in Google Play within half a day or so.

Summary

Woohoo! We built an app from *zero to published* in both stores, Apple App Store and Google Play. This has been quite an adventure with many twists and turns. We sincerely hope that this has given you deep insight into NativeScript for Angular app development as well as demystified any areas of this exciting tech stack for those who have been curious.

Both NativeScript and Angular have thriving global communities, and we encourage you to get involved, speak about your experiences, and share your excitement with others with all the exhilarating projects you and your teams may be working on. Never hesitate to reach out and ask for help, as we all take responsibility in our love and admiration for these two technologies.

There are some additional helpful resources that you can check out:

- `http://forum.nativescript.org`
- `http://nativescript.rocks`

And of course get to know the docs!
`http://docs.nativescript.org/angular/start/introduction.html`

Cheers!

Index

B

C

U

Y

Z

www.ingramcontent.com/pod-product-compliance
Lightning Source LLC
LaVergne TN
LVHW081513050326
832903LV00025B/1467